MORAL THEORY AND CLIMATE CHANGE

Climate change has become the most pressing moral and political problem of our time. Ethical theories help us think clearly and more fully about important moral and political issues. And yet, to date, there have been no books that have brought together a broad range of ethical theories to apply them systematically to the problems of climate change. This volume fills that deep need. Two preliminary chapters—an up-to-date synopsis of climate science and an overview of the ethical issues raised by climate change—set the stage. After this, ten leading ethicists in ten separate chapters each present a major ethical theory (or, more broadly, perspective) and discuss the implications of that view for how we decide to respond to a rapidly warming planet. Each chapter first provides a brief exposition of the view before working out what that theory "has to say" about climate change and our response to the problems it poses.

Key features:

- Up-to-date synopsis of climate science
- Clear overviews of a wide range of ethical theories and perspectives by leading experts
- Insightful discussions of the implications of these theories and perspectives for our response to climate change
- A unique opportunity to assess the relative strengths and weaknesses of various ethical viewpoints.

Dale E. Miller is a Professor of Philosophy at Old Dominion University and the editor-in-chief of *Utilitas*. He is the author of *J. S. Mill: Moral, Social and Political Thought* (2010) and he is a co-editor of *Rules, Morality, and Consequences* (2000); *John Stuart Mill and the Art of Life* (2011); *The Cambridge Companion to Utilitarianism* (2014); and *A Companion to Mill* (2017).

Ben Eggleston is a Professor of Philosophy at the University of Kansas. He is the editor of *Utilitarianism: With Related Remarks from Mill's Other Writings*, by John Stuart Mill (2017), and a co-editor of *John Stuart Mill and the Art of Life* (2011) and *The Cambridge Companion to Utilitarianism* (2014). His research interests include normative ethics, rational choice theory, and philosophy of law.

MORAL THEORY AND CLIMATE CHANGE

Ethical Perspectives on a Warming Planet

Edited by Dale E. Miller and Ben Eggleston

Routledge
Taylor & Francis Group
NEW YORK AND LONDON

First published 2020
by Routledge
52 Vanderbilt Avenue, New York, NY 10017

and by Routledge
2 Park Square, Milton Park, Abingdon, Oxon, OX14 4RN

Routledge is an imprint of the Taylor & Francis Group, an informa business

Library of Congress Cataloging-in-Publication Data
A catalog record for this title has been requested

ISBN: 978-1-138-70000-0 (hbk)
ISBN: 978-1-138-67827-9 (pbk)
ISBN: 978-1-315-20506-9 (ebk)

Typeset in Bembo
by Newgen Publishing UK

CONTENTS

FIGURES AND TABLES

Figures

Tables

CONTRIBUTORS

Sophie Grace Chappell is Professor of Philosophy at the Open University, UK. Her books include *Aristotle and Augustine on Freedom* (Macmillan, 1995), *Understanding Human Goods* (Edinburgh, 2003), *The Inescapable Self: An Introduction to Philosophy* (Orion, 2005), *Reading Plato's Theaetetus* (Hackett, 2005), *Ethics and Experience* (Acumen, 2009), and *Knowing What to Do: Imagination, Virtue, and Platonism in Ethics* (Oxford University Press, 2014).

Paul Clements is a Professor of Political Science at Western Michigan University where he directs the Master of International Development Administration program. His publications include *Rawlsian Political Analysis: Rethinking the Microfoundations of Social Science* with Notre Dame University Press, "A Rawlsian Analysis of the Plight of Bihar" in *Studies in Comparative International Development*, "Rawlsian Ethics of Climate Change" in *Critical Criminology*, and several articles and book chapters about international financial institutions and evaluation of foreign aid. He is a co-founder of Western Michigan University's Climate Change Working Group.

Regina Cochrane is an associate professor in the Department of Political Science at the University of Calgary. She completed a PhD in political theory at York University (Toronto) and a SSHRC post-doctoral fellowship in political philosophy at the University of Toronto. Her area of specialization is contemporary political theory, especially feminist/green political theory, social and global justice, Adorno versus counter-Enlightenment perspectives, and international development theory.

William Edelglass is Professor of Philosophy and Environmental Studies at Marlboro College and Director of Studies at the Barre Center for Buddhist

Studies. His research is primarily in the areas of Buddhist philosophy, environmental philosophy, and 20th-century French and German thought. William is chair of the Board of Directors of the International Association of Environmental Philosophy and is co-editor of the journal *Environmental Philosophy*. He is also co-editor of *Buddhist Philosophy: Essential Readings*, the *Oxford Handbook of World Philosophy*, and *Facing Nature: Levinas and Environmental Thought*. William lives with his wife and two daughters on an off-the-grid homestead in southern Vermont.

Ben Eggleston is Professor of Philosophy at the University of Kansas. He is the editor of *Utilitarianism: With Related Remarks from Mill's Other Writings*, by John Stuart Mill (Hackett, 2017), and a co-editor of *John Stuart Mill and the Art of Life* (Oxford UP, 2011) and *The Cambridge Companion to Utilitarianism* (Cambridge UP, 2014). His research interests include normative ethics, rational choice theory, and philosophy of law.

Steven Fesmire is Professor of Philosophy at Radford University. He is editor of *The Oxford Handbook of Dewey* (Oxford University Press, 2020) and the author of *Dewey* (Routledge Press, 2015), winner of a 2015 *Choice* "Outstanding Academic Title" award. He is also the author of *John Dewey and Moral Imagination: Pragmatism in Ethics* (Indiana University Press, 2003), winner of a 2005 *Choice* "Outstanding Academic Title" award. He was a 2009 Fulbright scholar at Kyoto University and Kobe University in Japan and a 2016 fellow at the Institute for Advanced Studies in the Humanities at the University of Edinburgh in Scotland.

Colleen McCluskey is Professor of Philosophy at Saint Louis University. She is the author of *Thomas Aquinas on Moral Wrongdoing* (2017) and co-author (with Rebecca Konyndyk DeYoung and Christina Van Dyke) of *Aquinas's Ethics: Metaphysical Foundations, Moral Theory, and Theological Context* (2009). Her research interests include medieval philosophy, feminist philosophy, and philosophy of race.

Dale E. Miller is a Professor of Philosophy at Old Dominion University and the editor-in-chief of *Utilitas*. His book *J. S. Mill: Moral, Social and Political Thought* was published by Polity in 2010, and he is a co-editor of *Rules, Morality, and Consequences* (Edinburgh, 2000); *John Stuart Mill and the Art of Life* (Oxford UP, 2011); *The Cambridge Companion to Utilitarianism* (Cambridge UP, 2014); and *A Companion to Mill* (Blackwell, 2017).

Michael Moehler is an Associate Professor in the Department of Philosophy at Virginia Tech and Director of its Program in Philosophy, Politics, and Economics. In addition, he is a Core Faculty Member of the ASPECT Program and an Affiliate Associate Professor in the Department of Economics at Virginia Tech. His main research interests lie in moral and political philosophy, with a specific focus on the contractarian tradition. He is author of *Minimal Morality: A Multilevel Social Contract Theory* (Oxford University Press, 2018).

Alice Pinheiro Walla is W1 Professor of Political Philosophy (Global Justice) at the Philosophy & Economics Program, Department of Philosophy, University of Bayreuth, Germany. Her research work combines Kant scholarship and contemporary political and legal philosophy to formulate a distinctive Kantian approach to questions of public relevance such as the territorial rights of states and groups, property rights, the justification of welfare rights and global duties.

Hans-Peter Plag is a Professor of Ocean, Earth & Atmospheric Sciences at Old Dominion University, where he studies societal and environmental sustainability, climate change, sea level rise, Earth system dynamics, solid Earth geophysics, and geodesy. He is the founding Director of the Mitigation and Adaptation Research Institute at ODU. He has led the development of a Conservation Leadership program at undergraduate and graduate level. He contributes to the Group on Earth Observations, which is implementing the Global Earth Observation System of Systems with particular focus on Earth observations in support of the UN's Sustainable Development Goals.

Madison Powers is Professor of Philosophy, Francis J. McNamara, Jr. Chair in Ethics, and Senior Research Scholar, Kennedy Institute of Ethics, Georgetown University. With Ruth Faden, he is co-author of *Social Justice: The Moral Foundations of Public Health and Health Policy* (Oxford University Press, 2006; rev. paperback edition, 2008) and *Structural Injustice: Power, Advantage, and Human Rights* (Oxford University Press, 2019).

INTRODUCTION

Dale E. Miller

With as much as has recently been written about climate ethics, this collection's approach to the topic is still quite novel. Two preliminary chapters—an up-to-date synopsis of climate science and an overview of the ethical issues raised by climate change—set the stage. After this, however, each remaining chapter considers what guidance a different ethical theory or (more broadly) perspective can offer us in coming to terms with these issues. Of course, it is not unusual for a philosopher to spend a few paragraphs comparing and contrasting the answers of two or three different theoretical perspectives to some moral quandary, and this is as common in the climate ethics literature as anywhere else. This collection is quite distinctive, however, in terms of the range of perspectives included, the detail in which their implications are drawn out, and the fact that each discussion is being conducted by an expert on the particular view in question—one whose sympathies lie with it.

This is not to say, though, that every "theory" chapter will follow the same template or speak to precisely the same points. Contributors were given some suggestions about issues in climate ethics that they might consider, e.g., what obligations that existing people have toward future people and how the burdens of mitigation and adaptation ought to be shared internationally. Part of what ethical theories differ about, though, is just which issues are most worth discussing. Contributors were left free to organize their chapters in whatever way would permit the most illuminating and fruitful presentations of their respective views.

This approach invites a certain criticism, which is that it reflects an objectionably "top-down" or "formalist" model of how to think about concrete ethical problems philosophically, one that involves starting with a moral principle or two, adding the relevant facts, and then mechanically deducing conclusions. The familiar term "applied ethics" is arguably an implicit affirmation of this model, and some prominent philosophers have explicitly endorsed it. K. Danner Clouser and

Bernard Gert, for example, observe that "The value of using a single unified moral theory to deal with the ethical issues that arise in medicine and all other fields, is that it provides a single clear, coherent, and comprehensive decision procedure for arriving at answers" (1990, 233). R. M. Hare goes even further, contending that the top-down model is the only rational alternative to philosophers' simply labelling any assertions that they cannot otherwise defend "intuitions" and treating them as foundations (1975). He takes putting this model into practice to be entirely straightforward. Hare's own theory is a version of act utilitarianism, and he confidently asserts that "in fact it is not difficult, if one gets the hang of the general theory, to apply it in particular fields like business and the professions. Ethics is ethics whatever one's vocation" (1992, 19).

Yet the top-down model has been vigorously challenged. Some philosophers signal their discomfort with it by using methodologically neutral terms like "practical ethics" in place of "applied ethics" (although many philosophers continue to use the more common term without meaning to suggest allegiance to this model). Michael D. Bayles offers a comprehensive list of objections to top-down ethical reasoning, which he calls "deductivism":

> First, [deductivism] leads to rigid and fixed distinctions and priorities that fail to aid in concrete decision making. … Second, there is no simple classification of facts that enables one simply to apply a rule or principle; facts do not come neatly labeled for the application of rules and principles. Third, deductivists are not responsive to the values intrinsic to the contexts in which problems arise. Their accounts of problems often ignore the wider cultural and historical context. … Fourth, deductivists assume that their moral theories are adequate, and consequently will not amend their theories, no matter how bizarre the results. … Fifth, deductivists assume that justification takes the form of logical deduction, of assuming cases under fixed rules. However, this model of reasoning has been strongly criticized in both philosophy of science and ethics
>
> *(Bayles 1984, 98; see also Beauchamp 2003, 8).*

Richard Norman further charges that the model is reductive, since "the conception of applied ethics as the application of … 'top-down theory'" involves artificially restricting the range of considerations that can count as justificatory practical reasons (2000).

Some of these objections presuppose that the top-down model is being implemented in an unnecessarily ham-handed way—which might, of course, frequently be the case—or that the theory being applied is itself unwarranted. Suppose that there *is* some "correct" ethical theory, whatever precisely correctness means on one's preferred metaethics—true, best justified, etc. This theory would not need to be amended in light of the results of applying it; if any of its implications seemed bizarre to us, then either we would have applied it improperly or our intuitions would be unreliable. Similarly, this theory would take account of as much cultural and historical context as is relevant (and no more). But other objections may cut more deeply and favor other models for thinking philosophically about concrete ethical problems.

The alternatives are various. There is a "bottom-up" model which starts with judgments about specific cases and which may take the form of casuistry, i.e., analogical reasoning from paradigm instances (Beauchamp 2003, 8–10; Paulo 2015). There is the "wide reflective equilibrium" approach according to which we should engage in both top-down and bottom-up thinking simultaneously, revising the commitments in which we have the least confidence until we have reached overall coherence in our ethical views (Beauchamp 2003, 10–12; Daniels 2018). And there is what Robert Baker and Laurence B. McCullough call the "appropriative" model, on which doing practical ethics "involves recontextualizing philosophical concepts and theory fragments to adapt them to practical purposes" (2007, 5).

This is not the place for a detailed comparison of the strengths and weaknesses of these different models. The only question here is whether the alleged deficiencies of the top-down approach to practical ethics model diminish the value of this collection, and they do not. Even if the top-down model should be abandoned, this would not rob this volume of interest.

Why not? One answer is that not all the viewpoints represented in this collection could serve as starting points for the top-down model, which requires that there be one or more general principles that can serve as the major premises of syllogisms. This is especially true of those perspectives that would not usually be classified as ethical theories, e.g., the ethical pragmatism that Steven Fesmire discusses or the Levinasian approach that is the subject of William Edelglass's chapter. But it is equally true of Sophie Grace Chappell's treatment of virtue ethics, despite the fact that this view is commonly referred to as a theory (and one that on other treatments might well lend itself to the top-down model).

Yet while this answer is true as far as it goes, it applies to only a portion of the viewpoints that are canvassed. Of the rest, it is admittedly the case not only that they could serve as starting points for the top-down deductive approach but also that this approach frequently figures in their respective chapters. Still, even someone who rejects this model should be able to see considerable value in these discussions. At least this is true as long as she does not reject the enterprise of ethical theorizing altogether. And while the list of alternative models of practical ethics that we offered above was not meant to be exhaustive, on each of these alternatives there is still some point to traditional philosophical work in ethical theory. On the appropriative model, those who work in practical ethics exploit "fragments" of ethical theories, and so there is use-value for them in seeing moral theories developed and their implications drawn out, even if they do not ultimately embrace any specific theory in its entirety. On the model of wide reflective equilibrium, top-down reasoning constitutes half of what "doing" practical ethics requires.

This leaves the bottom-up model, on which doing practical ethics involves starting with relatively concrete and specific ethical judgments and then moving toward greater abstraction and generality. The "narrow" version of reflective equilibrium is an example of this model. It might not be immediately obvious that this volume will hold much interest for the adherents of the bottom-up approach, insofar as most of the discussion in most chapters moves in the opposite direction. However, extracting ethical principles from intuitions or "considered moral judgments" is not necessarily a straightforward exercise; it is not merely a matter of drawing a curve that fits a number of dots on a scatter diagram. Work that draws out the concrete and specific implications of different familiar sets of ethical principles could help to guide someone who intends to move in the opposite direction, insofar as it shows which principles cohere with which particular judgments. In fact, there's nothing in the bottom-up model that would preclude someone from learning that a given theory has implications that they find very plausible for a wide range of issues—among which environmental issues in general and global warming in particular might be included—and subscribing to it on this basis. Even if this would mean that in some sense their thinking started with the theory and moved down, they would still see justification as moving from the bottom up.

In sum, while one will encounter some examples of top-down ethical reasoning in what follows, it would be a mistake to conclude that individual authors or the editors take this to be the one proper way to approach practical ethics generally or climate ethics specifically. The takeaway is instead only that this sort of reasoning, when done well, can at least make a valuable contribution to a larger conversation.

This collection's value extends beyond the contribution that it makes to helping us think through our response to the threat of climate change. It also makes a contribution to the long ongoing conversation about the strengths and weaknesses of different ethical perspectives. Even if there is more to doing practical ethics than applying abstract theories, seeing how multiple theories could be applied to the same practical problem does a great deal to illuminate their similarities and differences, their respective strengths and weaknesses. It is hard to imagine a better way for students who are relatively new to the study of ethics to learn about the wide range of ethical perspectives that we have surveyed than to compare and contrast what they call for in practice, and every chapter in the volume should be accessible to motivated undergraduates.

But trained philosophers working in ethics will also find much to interest them in this collection, whether they specialize in climate ethics or not. First, it is unlikely that many moral philosophers have more than a passing acquaintance with the full range of perspectives that are included. Those whose training was broadly "analytic," for example, may be barely acquainted with Levinas. (This was certainly true of the editors.) Second, in several chapters they will find theoretical innovations, or at least discussions of some previously unexplored possibilities afforded by the different perspectives. Finally, third, they may find that climate change represents a unique test of the adequacy of ethical outlooks. Part of what is distinctive about the challenge posed by climate change is that it is the product of many actions by people around the world (albeit in some parts of the world far more than others) that considered in and of themselves may seem too innocent or at least too ordinary to merit condemnation. Now suppose that some particular ethical perspective came to be generally adopted very quickly, globally, at least to the extent that more or less everyone acts as it says that they should. Would this result in humanity's largely averting the ills of climate change, insofar as it is still in our power to do so? If the answer to this question is "no," for some given perspective, then arguably this is a serious deficiency—perhaps a disqualifying one. This collection offers philosophers an opportunity to judge how well a wide range of perspectives rise to meet this test.

Bibliography

Baker, Robert and Laurence B. McCullough. 2007. "Medical Ethics' Appropriation of Moral Philosophy: The Case of the Sympathetic and the Unsympathetic Physician." *Kennedy Institute of Ethics Journal* 17 (1): 3–22.

Bayles, M. D. 1984. "Moral Theory and Application." *Social Theory and Practice* 10 (1): 97–120.

Beauchamp, Tom K. 2003. "The Nature of Applied Ethics." In *A Companion to Applied Ethics*, edited by R. G. Frey and Christopher Heath Wellman. Malden MA: Blackwell.

Clouser, K. Danner and Bernard Gert. 1990. "A Critique of Principalism." *The Journal of Medicine and Philosophy* 15 (2): 219–36.

Daniels, Norman. 2018. "Reflective Equilibrium." *The Stanford Encyclopedia of Philosophy* (Fall 2018 Edition), edited by Edward N. Zalta. https://plato.stanford.edu/archives/fall2018/entries/reflective-equilibrium/.

Hare, R. M. 1975. "Abortion and the Golden Rule." *Philosophy and Public Affairs* 4 (3): 201–22.

———. 1992. "One Philosopher's Approach to Business and Professional Ethics." *Business and Professional Ethics Journal* 11 (2): 3–19.

Norman, Richard. 2000. "Applied Ethics: What is Applied to What?" *Utilitas* 12 (2): 119–36.

Paulo, Norbert. 2015. "Casuistry as Common Law Morality." *Theoretical Medicine and Bioethics* 36 (6): 373–89.

1

MODERN CLIMATE CHANGE

A Symptom of a Single-Species High-Energy Pulse

Hans-Peter Plag

Introduction

In 1912, Francis Molena contemplated the possibility of human activities causing climate change, and he concluded:

> It is largely the courageous, enterprising, and ingenious American whose brains are changing the world. Yet even the dull foreigner, who burrows in the earth by the faint gleam of his miner's lamp, not only supports his family and helps to feed the consuming furnaces of modern industry, but by his toil in the dirt and darkness adds to the carbon dioxide in the earth's atmosphere so that men in generations to come shall enjoy milder breezes and live under sunnier skies.
>
> *Molena, 1912*

Today, more than a century later, atmospheric CO_2 is much higher than Molena expected, and instead of enjoying the "milder breezes," fear of stronger storms is increasing, while the "sunnier skies" are associated with more frequent deadly heat waves and prolonged droughts. During the last century and particularly the last several decades, climate has been changing at a much higher rate than before 1900 throughout the Holocene, the last geological epoch that started approximately 12,000 years ago (e.g., Gaffney & Steffen, 2017). Evidence documenting modern climate change is abundant (e.g., Pachauri et al., 2014; IPCC, 2018) and the observed changes in the chemistry of the climate system can be linked to human activities. However, recent impacts of modern society on the planetary system go far beyond changing the climate. The rapid modern climate change is part of a larger syndrome of accelerated modern global change and can only be characterized and understood as part of this syndrome.

The planetary system can be viewed as a life-support system for a very large number of fine-tuned subsystems of species interacting with each other within this system and slowly changing it. The concept of a planetary life-support system of systems has been utilized recently in scientific assessments of the interaction of humanity with the planet (e.g., Young & Steffen, 2009; Pearce, 2010) and in communications of the anthropogenic degradation of the system to the public and world leaders (e.g., Wallström et al., 2004; Barnosky et al., 2014).

The concept of the Earth's life-support system was used by Griggs et al. (2013) to operationalize the principle of sustainable development by defining sustainable development as "a development that meets the needs of the present while safe-guarding the Earth's life-support system on which the welfare of current and future generations depends." Communities of human and non-human animals embedded in the planetary system interact with this life-support system through flows of matter and energy (Figure 1.1). In many aspects, the Earth's life-support system is similar to the organism of an animal, in which many fine-tuned processes generate flows that keep the organism in homeostasis. Like the planetary system, the animal organism provides a life-support system for many other organisms that are crucial for the generation of the flows that maintain homeostasis. Central to the

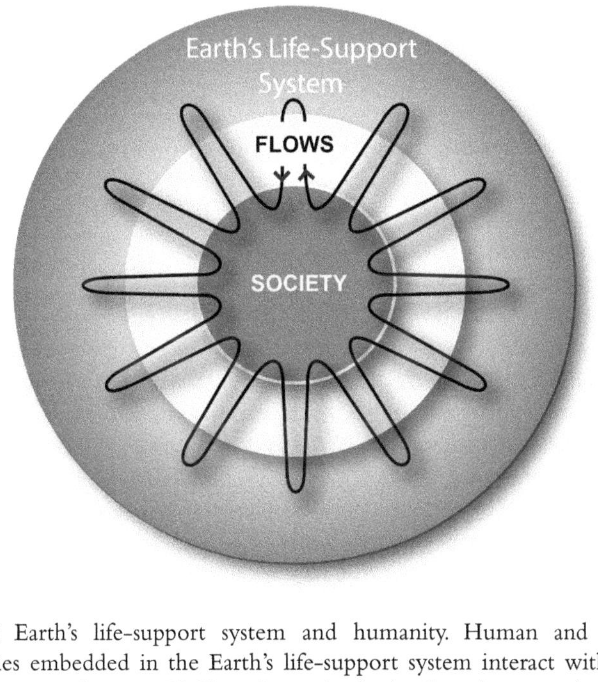

FIGURE 1.1 Earth's life-support system and humanity. Human and non-human communities embedded in the Earth's life-support system interact with it through flows of energy and matter. Unlike other animals, the flows between the life-support system and human communities are regulated by ethics, social norms, and the mainstream economic model.

Source: Modified from Plag & Jules-Plag (2017).

understanding of the physiological functioning of a body system is the integrated nature of chemistry and physics, coordinated homeostatic control mechanisms, and continuous communication between cells (Widmaier et al., 2016). Central to the understanding of the functioning of the Earth's life-support system is the integrated nature of chemistry and physics and the coordinated functioning of homeostatic control mechanisms provided by the continuous interaction of the "web of life" (Capra, 1996) embedded in, and integral to, the planetary life-support system.

Life has impacted the physiology of the planetary system from the start by creating and changing flows. To some extent, life determined many chemical and physical system variables throughout time and impacted climate. The flows manipulated by life kept the system in homeostasis and provided long stable states for life to evolve and occupy large regions of the planetary surface layers. Changes in the flows were slow and allowed species to adapt to equally slow changes in the mean state of the system, as well as most of the fluctuations around the mean.

In the case of human communities, the flows between the Earth's life-support system and society are regulated not only by human needs but also by ethics, social norms, and economic rules and practices. The modern growth-dependent economy has facilitated growth by increasing most of the flows by several orders of magnitude, and these rapid changes in the planetary physiology have resulted in major changes in the biological, chemical, and physical conditions in the Earth's life-support system. Among them is modern climate change.

Fever in a homeothermic species is a symptom of a malfunctioning of that system and often part of a syndrome of changes in the organism resulting from disruptions in the physiology of the organism. These disruptions can result, for example, from a breakdown of homeostatic control mechanisms, alterations of flows, or the attack of viruses or bacteria. The effectiveness of an external cure depends upon an accurate diagnosis of the disruption.

Similarly, a rapid increase in the global ocean and air temperature, which is indicating a rapid increase in the energy stored in the coupled atmosphere-ocean system, is a result of a disruption in the planetary physiology. The symptom of global warming indicates that the homeothermic processes in the Earth's life-support system no longer are functioning to keep the system in homeostasis. Consequently, only by considering the full planetary physiology can the symptoms of global warming and modern climate change be fully understood and traced back to the underlying cause, the "sickness."

An alien outside observer would not characterize the syndrome as anthropogenic but rather aim to see it as a distortion in the planetary life-support system that could have come from any individual or group of species. Having access to the very large database humans have compiled in the last few centuries, the alien observer would see the obvious: The distorting modern global change is the result of a single species that in a very short time of less than 200 years released a very large pulse of energy into the environment by tapping into energy resources stored in the planetary system over hundreds of millions of years. This species is using this energy pulse to re-engineer the planetary physiology by changing its chemical and physical state and modifying crucial flows by several orders of magnitude. To capture the main aspects of this distortion, the alien might denote it as a "single-species high-energy pulse" syndrome. The species causing the pulse acts as a virus in the Earth's life-support system, a virus that found the means to change almost all flows to sustain its population growth at an unprecedented rate, invasively occupy all regions of the planet's surface, and eliminate many other potentially competing species. By doing so, it destabilized the homeostatic mechanisms and initiated a rapid transition of the planetary system towards a new and currently unknown homeostasis.

Scientific evidence pointing to an ecological breakdown has become ubiquitous, and concerned scientists have issued warnings to humanity (Union of Concerned Scientists, 1992; Ripple et al., 2017). Major newspapers, including *The Guardian* and the *New York Times*, have picked up the alarming scientific findings concerning soil depletion, deforestation, and the collapse of fish stocks and insect populations. There is a growing consensus that these crises are driven by an economic model focused on unlimited growth of production and consumption, which is fundamentally reorganizing the flows in the Earth's life-support system.

Modern climate change is extensively documented in scientific literature and assessment reports at national, regional, and intergovernmental levels, and there is little benefit in adding another summary here. However, considering modern climate change as a symptom of the larger single-species high-energy pulse syndrome is a novel contribution. Looking at the Earth's life-support system from a medical point of view to assess its health, these five questions seem crucial:

- What is the baseline for a healthy planetary life-support system and what are the normal ranges of essential variables of this system?
- What characterizes the syndrome of the single-species high-energy pulse that the system is showing recently?
- What is the diagnosis of the underlying cause of this syndrome?
- What foresight can be developed with respect to the full spectrum of possible futures of the system and what are the likelihoods of these futures to realize?
- Is there a therapy to address the cause of the syndrome?

Understanding the extent of modern global change requires a baseline against which the changes can be characterized.

The remainder of this chapter is organized as follows: In the second section, selected aspects of the planetary baseline for a stable life-support system will be considered, including the background ranges of climate variability and change. In the third section, the full extent of the syndrome of modern global change is assessed against this baseline. The context of Earth's life-support system allows for a diagnosis of the main cause of this syndrome and this is discussed in the fourth section. Foresight about the full spectrum of possible futures of the planetary life-support system and humanity embedded in it is needed to inform our actions, and this foresight can be developed by considering several scenarios of human interactions with the planetary system, as discussed in the fifth section. The sixth section considers a therapy that directly results from the diagnosis of the cause of modern global change. Finally, concluding remarks are based on the working hypothesis that the future does not exist and has to be created. Since humanity has engaged in re-engineering and operating the planetary system, humanity also has responsibility to create a future that is consistent with the ethics it promotes.

Baseline and Normal Ranges

Climate Basics

The Earth's climate is the result of three main factors: the incoming solar radiation (solar irradiance), which determines the energy input to the Earth system; the albedo of the Earth's surface, which determines the radiation reflected at the Earth's surface; and the amount of so-called greenhouse gases in the atmosphere, which determines how much of the long-wavelength radiation is captured by the atmosphere. If any of these three factors change, climate will change. However, the spatial and temporal patterns of climate change can be complex and difficult to attribute to specific changes in any of the three forcing factors. The amount of heat stored or released in response to changes in the three forcing factors depends on the heat capacity of the main components, including the atmosphere, ocean, cryosphere, and solid land surface. Changes in the overall heat stored in the different components of the climate system have an impact on the dynamics within the climate system, in particular, the ocean and atmospheric circulation and the flows in the associated water cycle.

Climate can change on a wide range of spatial and temporal scales, and the changes in climate variables such as surface air temperature, air pressure, wind fields, air and soil moisture, precipitation, and evapotranspiration, among others, can differ along these scales. For example, the changes in five-year averaged air temperature between 1900 and 2010 vary spatially between -2°C and +4°C, with the increases in the Arctic on average being much larger than in other geographical regions. Over time, global mean air temperature has varied over a total range of about 5°C over the last 800,000 years (e.g., Hansen et al., 2008), but changes on century time scales have been very small (normally much less than 0.5°C).

Solar irradiance exhibits small fluctuations over time. Most remarkable for human time scales is the solar sunspot cycle of approximately 11 years. The corresponding climate variations are minuscule, and the processes that cause the variations are still being researched (Rind et al., 2008).

The Earth's albedo, i.e., the ratio of reflected to incoming radiation, depends on the Earth's surface properties. Ice and snow-covered areas reflect most of the solar radiations, while open water and land areas reflect much less. Urban areas also have a lower albedo than, for example, grasslands and forests.

Without an atmosphere, the mean global temperature on the Earth's surface would be close to -18°C, and this temperature would only depend on the incoming solar radiation. With the atmosphere present, the global mean surface temperature is today close to 16°C, and this much higher temperature is mainly the result of the greenhouse gases in the atmosphere. The greenhouse gases most abundant in the atmosphere include H_2O, CO_2, CH_4, and N_2O, while other gases such as fluorinated gases have high greenhouse potential but are less abundant in the atmosphere. If the atmospheric concentrations of greenhouse gases increase, then the surface air temperature also increases (Arrhenius, 1896).

However, due to the presence of the ocean, the full effect of a change in greenhouse gases can take a long time to develop. Water has a specific heat capacity more than 3,000 times that of air, and the slow heating of the ocean thus delays the atmosphere in reaching the equilibrium temperature consistent with a given level of atmospheric greenhouse gases. Thus, the climate system reacts to changes in greenhouse gases not like a greenhouse but rather like a pool house, in which the swimming pool provides a large thermal mass that delays changes in the air temperature.

The warming ocean also contributes to changes in the climate dynamics. A warmer ocean evaporates more water, and the amount of water vapor that can be stored in air increases strongly with the air temperature. For example, at 40°C, air can hold almost 5.5 times the water it can hold at 10°C. Thus, warmer air can store much more water before condensation and precipitation happen, which results in longer intervals without precipitation followed by more extreme precipitation events. The water vapor represents latent heat, that is, energy that becomes available again when the water vapor condensates. This energy is available to drive storms. The warmer the air, the more latent heat it can hold and make available for storms once condensation is triggered.

Thus climate change is the result of changes in flows: the flow of solar energy into the atmosphere-oceans system and storage of heat mainly in the ocean; the flow of water in the ocean currents that distributes energy throughout the ocean; the flow of energy in the form of latent heat carried by water vapor from the oceans to the atmosphere; the transformation of latent heat into kinetic energy when the water vapor condensates; the flow of air in storms that can bring devastation to human and non-human communities; and the flow of water in the global water cycle that can cause floods and droughts.

Normal Ranges

The Earth's life-support system is a dynamic system and changes are continuously taking place at a wide range of spatial and temporal scales. However, there are many biological, chemical, and physical feedback processes and loops that create prolonged periods of homeostatic stability, in which the system exhibits relatively small variations around a surprisingly stable mean state. For the last 1 million years, the climate system has fluctuated between cold glacial periods (ice ages) and warm inter-glacial periods, with the transition from warm to cold periods often taking much longer than the reverse transition. Data gathered from many natural recorders of climate variability, e.g., tree rings, ice cores, fossil pollen, ocean sediments, coral, and historical data, has been used to reconstruct many variables of the climate system and the overall planetary system. This so-called paleo-data can be used to establish a baseline indicating the range in which these variables have fluctuated during the last approximately 1 million years. This baseline provides a basis for the assessment of the syndrome of modern global change.

An essential variable associated with the planetary energy state is the Earth's energy imbalance. This is the difference between incoming solar energy and outgoing energy. Among others, the imbalance results from solar energy being used to produce biomaterial that partly is being deposited in subsurface reservoirs as fossil fuels. The energy used by non-human species, including plants, to generate and modify flows is at a very low level compared with the incoming solar energy. As a result, averaged over the last 200 million years, the imbalance was roughly on the order of 10^{-10}; that is, on the order of 10 megawatts of the incoming roughly 98,000 terawatts of solar energy were not returned to space. On shorter time scales, the energy imbalance varied between much larger positive and negative values particularly during the cooling or warming of ocean waters and the aggregation and melting of large ice sheets.

An essential variable indicating the stability of the biological system is the rate at which species go extinct. Although it is difficult to estimate background rates for extinction (Pimm et al., 2014; Ceballos et al., 2015), it appears that outside the five known periods of mass extinction (e.g., Bond & Grasby, 2017; Starr, 2018) the extinction rates are very low. For mammals, a recent study estimates the background rate to be two extinctions per 10,000 species per 100 years (Ceballos et al., 2015, and the references therein). However, during the transitions between the glacial and inter-glacial periods in the last 1 million years, extinction rates appear to have been higher. For example, terrestrial ecosystems experienced major transformations during the most recent warming and associated climate change after the last ice age, and many plant species became extinct (Nolan et al., 2018).

For the climate system, a number of physical and chemical variables are essential (Bojinski et al., 2014). Table 1.1 gives the ranges for a few essential climate variables for the last 800,000 years and for the Holocene. The total ranges for the

TABLE 1.1 "Normal range" of selected essential climate variables.

Variable	*Long-term range*	*Holocene*	*Current (Jan. 2018)*	*Projected 2100*
CO_2	170–300 ppm	270–285 ppm	407.4 ppm	500–900 ppm
CH_4	320–800 ppb	550–800 ppb	1855 ppb	2000–2500 ppb
Greenhouse gas forcing	-4.0–0.0 W/m^2	-0.3–0.0 W/m^2	1.0 W/m^2	3.5–5.0 W/m^2
Albedo forcing	-4.0–0.0 W/m^2	-0.2–0.0 W/m^2	0.3 W/m^2	1.5–2.5 W/m^2
Global mean surface temperature	-4.0–1.0°C	0.0–1.0°C	1.0°C	2–6°C
Global mean sea level	-130–5 m	-3–0.0 m*	0.3 m	1–5 m

*This is for the last 7,000 years during the time sea level was exceptionally stable.
The normal ranges for the selected variables are determined based on data provided by Hansen et al. (2008) for the last 800,000 years. The long-term range is for the last 800,000 years prior to industrialization. The range for the Holocene is for the last 12,000 years before industrialization. Current values are for 2017. The projected ranges are those published based on model studies.

greenhouse gases CO_2 and CH_4 are 130 ppm and 480 ppb, respectively, translating into a total range of the greenhouse gas forcing of 4 W/m^2, which is similar to the total range of albedo forcing. Using the ranges in Table 1.1, 1 ppm change in atmospheric CO_2 corresponds to a change of 1 m in global mean sea level, and 25 ppm corresponds to 1°C in global mean air temperature.

During the Holocene, the range for the greenhouse gases was only a fraction of the range during the last 800,000. Likewise, global mean temperature variations were limited to a range of 1°C. During the first part of the Holocene, sea level continued to rise nearly 50 m until an equilibrium between the air temperature and land-based ice masses was reached about 7,000 years ago. Since then, the range of global sea level change was only on the order of a few meters.

The extreme stability of climate during the Holocene provided a safe operating space for humanity (Rockström et al., 2009), in which modern society could emerge. When climate became stable at the beginning of the Holocene, the transition to agriculture was possible. When sea level became stable, permanent settlements became feasible in river deltas, which provided valuable ecosystem services and logistical advantages, and cities could develop.

The Holocene can be characterized by planetary boundaries for essential variables. Rockström et al. (2009) identified boundaries for nine variables: biogeochemical flows (in particular nitrogen and phosphorus cycles), global freshwater, change in land use, biodiversity loss (extinction rate), atmospheric aerosol loading, chemical pollution (and new constituents), climate change, ocean acidification, and stratospheric ozone depletion. They found that several of these boundaries have been crossed recently due to human activities, and this is pushing the Earth's life-support system out of the safe operating space.

The Single-Species High-Energy Pulse Syndrome

Climate Symptoms

The observational evidence of a global temperature rise of close to 1°C since 1880 is overwhelming (Pachauri et al., 2014; IPCC, 2018) and indisputable. There are large spatial and temporal variations in the increase. For example, a rapid warming took place in the Arctic, where the land surface temperature increased by 3.5°C since 1900 (Morrison, 2017).

The most fundamental variable reflecting the extent of global warming is the Earth's energy imbalance. In recent decades, this imbalance has increased by many orders of magnitude (Hansen et al., 2011; Trenberth et al., 2014; von Schuckmann et al., 2016; Cheng et al., 2019). During the first decade of the 21st century the imbalance was estimated at 250–500 terawatts (Trenberth et al., 2014). Most of this energy is being stored in the ocean (> 90%) and only 1% is stored in the atmosphere and causing the increase in global air temperature (Laffoley & Baxter, 2016, and the reference therein). The remaining fractions go into melting of ice and the warming of the land surface. The energy currently stored in the planetary climate system is 14 to 28 times the total energy used by humanity. Thus, the anthropogenic re-engineering of the planetary system has an amplified impact on the energy imbalance of the planet, which is now on the order of 1 million to 10 million times larger than the long-term pre-human background rate (see p. 000). The energy stored leads to significant changes in land and sea-ice cover, snow cover, and increased desertification, which all impact the albedo of Earth's surface. However, the full impact of the additional heat on global mean air temperature is not yet being felt because of the large time-lag caused by the thermal mass of the ocean.

The many symptoms and full scale of modern climate change are described in a number of assessments at subnational (e.g., Bedsworth et al., 2018), national (e.g., Wuebbles et al., 2017), and international levels (e.g., Pachauri et al., 2014; IPCC, 2018). Atmospheric CO_2 has reached permanent levels of 405 ppm, i.e., more than 100 ppm above the baseline range. Given that during the last 800,000 years 25 ppm of CO_2 corresponded to 1°C in global mean air temperature, the current increase of only about 1°C above the pre-industrial temperature is most likely not reflecting the equilibrium temperature corresponding to a 100 ppm increase in CO_2. Moreover, several feedback loops seem to have been triggered, with the melting of permafrost having a high potential of contributing additional greenhouse gas emissions (e.g., Drake et al., 2015; Kohnert et al., 2017).

A few additional examples are considered here to illustrate the range of symptoms. On a regional scale, air temperatures have been found to increase faster during periods of droughts than on average: for example, in the southern and northeastern United States, concurrent occurrences of droughts and heat waves have caused compounding ecosystem and societal stresses (Chiang et al., 2018). For warmer air, much more water vapor is needed to reach the condensation point (p. 000), and when this point is reached, the resulting precipitation is much larger. Therefore, a warmer world will likely have many more extreme precipitation events. This is already visible in the amount of record flooding taking place in recent years in many regions (e.g., EASAC, 2013, 2018). The warming ocean and atmosphere appear to have increased the size and maximum precipitation of the 2018 Hurricane Florence compared with a similar hurricane under pre-warming conditions (Reed et al., 2018).

Many of the recent changes appear to have taken place earlier than predicted a few years ago. For example, the reduction in summer sea ice in the Arctic is taking place much faster than predicted, most likely due to underestimation of the increased summer solar heat absorption by the ice-free surface waters (Timmermans et al., 2018), which in turn accelerates the disappearance of sea ice in the region beyond the summer season. Importantly, a potential slow-down of the northward flow of warm surface waters and the matching southward flow of cooler waters at greater depths would have severe impacts for the climate on the Northern hemisphere, and it was not expected in the 21st century (National Research Council, 2013). However, there is now evidence that it has started to slow down (Bryden et al., 2005) and is slower than at any time in the last 1,600 years (Caesar et al., 2018; Thornalley et al., 2018).

Other Symptoms and Impacts

Like many other animal species, *Homo sapiens* has changed the physiology of Earth's life-support system since humans started to build simple dwellings more than 500,000 years ago. With the transition to agriculture, they started to control the flow of water (Bishop et al., 2017) and changed other flows including soil erosion, nutrient flows, and flows of bacteria. Already several thousand years ago, the use of fire and other means to clear forested areas for agricultural use changed the flow of carbon and started to increase atmospheric CO_2 with a likely impact on the long-term climate trend throughout the Holocene (Ruddiman, 2005).

In more recent times, the re-engineering of the land surface created many new flows and modified or interrupted existing flows. Destruction and degradation of ecosystems are the primary causes of extinction and the increasingly rapid declines in global biodiversity (Pereira et al., 2010). The fragmentation of habitats into smaller, isolated ecosystems separated by a matrix of built environment and human-transformed land cover leads to a greater exposure to anthropogenic flows along fragment edges, and this in turn initiates long-term changes in the structure and function of the remaining fragments (Haddad et al., 2015). Globally, 70% of the remaining forests are within 1 km of edges due to fragmentation, and this reduces biodiversity and impairs key ecosystem functions by decreasing biomass and altering flows in nutrient cycles (Haddad et al., 2015).

New boundaries in the aquatic system with steep gradients introduced by canals, dammed reservoirs, irrigation ditches, and pollution are changing species diversity, microbial communities, and nutrient levels in aquatic zones across the planet resulting in significant detrimental impacts on aquatic ecosystems (Bianchi & Morrison, 2018). The accelerated flows of nutrients into the ocean have created large dead zones and contribute to declining oxygen in the global ocean and coastal waters (Breitburg et al., 2018). Global ocean warming has a severe impact on marine ecosystems and is one of the main causes of widespread coral bleaching (Hughes et al., 2018; Frade et al., 2018). Ocean acidification caused by increased absorption of atmospheric CO_2 in the ocean adds another major threat: a threshold in the total carbon load added to the ocean that would lead to the catastrophic elimination of all marine life could be reached as early as 2050 to 2150 (Rothman, 2017).

Modern global change caused by human activities has had a significant impact on the biomass distribution. Over the last 5,000 years human activities have reduced total global biomass by about 50% (Smil, 2011), and this decline through harvesting of the wild and habitat conversion continues. Between 1900 and 2000, the carbon stored in humans increased by more than 400%, while for wild terrestrial animals it was reduced by 50%. Humans also increased the biomass of domesticated animals by 350%, with cattle accounting for two-thirds of the total mass of these animals (Smil, 2011). Besides a major impact on wildlife, the increase in cattle also creates a major source of greenhouse gases in the form of CH_4.

Modern climate change is also impacting the frequency and magnitude of other natural hazards. For example, the increased melting of glaciers and ice sheets has increased the risk of landslides and tsunamis triggered by these landslides (e.g., Higman et al., 2018). During the deglaciation that ended the last ice age, several megatsunamis were caused by subaerial landslides of sediments deposited around the melting ice sheets (Harbitz et al., 2014). It can be expected that the possible melting of the Greenland and Antarctic ice sheets will also create large sediment deposits on the surrounding shelves that will increase the probability of large subaerial landslides and megatsunamis.

In summary, the large range of symptoms within the single-species high-energy pulse syndrome indicate that the Earth's life-support system is rapidly moving out of the safe operating space for humanity. At least three of the nine planetary boundaries of this space have been crossed (biogeochemical cycles, extinction rates, climate change; Rockström et al., 2009; Steffen et al., 2015).

Diagnosis

Breaking Scaling Laws: The Journey to the Single-Species High Energy Pulse

Distortions in the planetary physiology have been caused by many events in Earth's long history. For example, episodes of high volcanic activity have impacted the chemical composition of the atmosphere and ocean, and in extreme cases

TABLE 1.2 Century-scale changes in essential variables of the Earth's life-support system.

Variable	*Before 1900*	*1900–2000*	*Acceleration*
Energy usage	0.01 TW/century	16 TW	1600
CO_2	0.2 ppm/century	120 ppm	600
Population	16 million/century	5.5 billion	350
Gini coefficient	0.003/century	0.3	100
Global mean temperature	0.01°C/century	1°C	100
Global mean sea level	0.05 m/century*	0.2 m	4

*Value is for the last 7,000 years, when sea level was exceptionally stable.
Changes per century in the essential variables of the climate system and humanity were very small up to 1900. The combination of technological innovations and humanity's newly gained access to seemingly unlimited energy facilitated in the last century changes in the planetary system that were several orders of magnitude larger than on average throughout the Holocene. A few variables such as global mean sea level, which lag behind in time, do not show the full impact yet.

caused mass extinction (Bond & Grasby, 2017). However, the changes in flows introduced by humanity in the high-energy pulse are exceptional in terms of the rate of change. Between 1900 and 2000, many essential variables exhibited changes that are 100 to 2,000 times larger than on average throughout the Holocene (Table 1.2). In particular, the rapid increase in energy usage during the last 100 years allowed the production of large amounts of fertilizers as well as a large-scale transformation of land to agricultural land. This resulted in an apparent transient increase of the planetary carrying capacity and sustained an increase of the global human population 350 times faster than before. The rapid increases of atmospheric CO_2 and global mean temperature are side effects of this dominance of *Homo sapiens*.

The minimum energy needed by animals (including humans) to stay alive is the basal metabolic rate, which can be expressed in watts. This basal metabolic rate depends on the mass of the animal, and the rates for different species are connected through a simple scaling law, which states that the basal metabolic rate is a constant times the mass of the animal to the power of 3/4. The constant differs among the three groups of homeotherms, poikilotherms, and unicellular organisms (West et al., 1997, and the references therein).

Homo sapiens is in the group of homeotherms, and the scaling law results in a range for the basal metabolic rate of 50 to 100 watts, depending on age, gender, and mass. Because they use fire to process food, humans can extract the energy to sustain this rate from food far more efficiently than most other species and they could afford to develop larger, energy-demanding brains (Herculano-Houzel, 2017). At the same time, it freed time that previously had been used to find and metabolically process the food. This put *Homo sapiens* on a distinctly different evolutionary path (Table 1.3). The use of fire for clearing forest and other land covers

TABLE 1.3 Humanity and energy.

Time	*Event*	*Impact*	*Population*	*Energy usage (terawatts)*
1,000,000 BP	Use of fire to prepare food	Allowed development of larger, more complex brains	?	0
10,000 BP	Transition to agriculture	Increased food security, reduced time needed to find food	4 million	< 0.000001
1000			50 million	< 0.001
1750	Advent of capitalism (Adam Smith) and colonialism	Industrialization and rapid acceleration of flows started; single-species dominance	650 million	0.1
1900		Threat of climate change and population growth identified	1.6 billion	0.5
1990	UNFCCC (1992) and 1st IPCC report (1994)	Climate change elevated to one of the main threats to humanity	5.3 billion	10.0
2000			6.0 billion	12.8
2010			6.9 billion	16.5
2100	Potential climate tragedy	Potential social collapse	1–14 billion	> 100?

The single-species high-energy pulse has its origin in humans learning to utilize fire to process food, which provided a basis for increasing the brain-body ratio. Later fire was used to clear spaces for agriculture. The focus on accumulation of human wealth in Europe, the disregard of Earth's life-support system, and the discounting of the future led to an unparalleled single-species dominance and rapid degradation of the life-support system for all.

for agriculture and pastures further eased food supply and freed time for other activities. The larger brain could be used to develop technologies to utilize more energy to transform the environment, triggering the development that eventually led to the single-species high-energy pulse; modern climate change is one of the most visible and severe symptoms of this syndrome.

Unlike other animals, today's *Homo sapiens* uses a large amount of energy to modify the planetary system and create conditions that are favorable and convenient for humans. Adding the energy used per capita to the basal metabolic rate, the resulting extended metabolic rate is much larger than what the scaling law indicates. In 2010, the global average energy used per capita was approximately 2,750 watts, resulting in an extended metabolic rate of 2,800 watts. Using the scaling law for homeotherms, an animal with a basal metabolic rate of 2,800 watts would have a mass of roughly 9,000 kg, which is equivalent to the mass of two large elephants. Based on the scaling law, humanity's energy usage is equivalent to what roughly 14 billion elephants would need. Thus, by breaking the scaling law, *Homo sapiens* became a planetary dominant force overloading the carrying capacity of Earth's life-support system.

Role of the Mainstream Economic Model

The current global mainstream economic model is based on a number of assumptions about the Earth's life-support system and humanity's relation to it, how economy works, and what the goal of economy is. The core purpose of today's economic model can be traced back to Adam Smith, who focused economy on the creation of human wealth (built capital, Smith, 1776). In the 18th century, the global population was less than 700 million and access to energy and technology was limited. Humanity was managing less than 1% of the Earth's surface. Built capital was the limiting factor, while natural capital was seemingly infinite. Consequently, it made sense to think of economy only in terms of marketed goods and services and to define the goal as increasing production and consumption without worrying much about impacts on the planetary life-support system.

The mainstream economic model does not consider the health of the planetary life-support system and the state of this system is not accounted for in any economic measure. While this approach allowed human economy to develop rapidly and facilitate an unparalleled growth of the global population, it required the acceleration of flows in the planetary physiology. As a result, the per capita use of materials, including metals, minerals, fossil fuels, and biomass, increased rapidly, particularly since 1950, and reached in 2008 on average 10 tons per person, i.e., 27 kg each day, or a global total of 68 gigatons of materials (Assadourian, 2013).

Economic growth has been directly linked to rapidly increasing energy use and greenhouse gas emissions. Global CO_2 emissions from fossil fuel combustion, cement production, and other industrial processes account for about 70% of total global greenhouse gas emissions (UNEP, 2017). Other emissions result from land use, land-use change, and forestry. Emissions are still increasing, although the rate of increases has been slowing down recently (UNEP, 2017). However, it is not clear whether this slowdown will continue as a result of a decoupling of CO_2 emissions and economic growth or reverse if economic growth is accelerating.

The apparent success of the current economic model in reducing poverty and increasing human welfare is based on a consumption of planetary non-renewable resources (Assadourian, 2013) and a rapid degradation of the planetary life-support system, including modern climate change. It is increasingly clear that the mainstream economic model is at the core of the current unsustainability. Only a reconceptualizing of the economic model can therefore provide a path to more sustainability (Costanza et al., 2013).

Foresight

Embedded in the Earth's life-support system, *Homo sapiens* today can be characterized as a virus in this system. The attempt to provide a prognosis of how this virus will impact the system's health is hampered by the unpredictability of the virus's actions. However, the spectrum of possible futures can be explored to inform foresight. In particular, models can be used to explore the range of possible futures of the climate system for a large set of scenarios for human-caused emissions of greenhouse gases. Unfortunately, the predictive capabilities of the models are limited because of epistemic uncertainties; that is, knowledge gaps. The importance of the knowledge gaps is emphasized by the many impacts of human activities on the planetary system that have not been predicted or even anticipated before they were discovered. For example, the large dead zones in the ocean caused by overloads of nutrients from agriculture, the rapid reduction in the area of sea ice in the Arctic, the increase in extinction rates, and the large change in the Earth's energy imbalance were either not predicted at all or underestimated by model prediction. However, the understanding of basic relationships often is sufficient to develop foresight in the sense of what are possible futures without predicting which future actually is going to realize. This foresight then allows a risk assessment that can inform decisions. In particular, complex systems stressed beyond their homeostatic limits will rather abruptly transition into new states that cannot be predicted. Mounting evidence suggests that in the case of the Earth's life-support system (including the climate system), the new states may well be inimical to humankind (e.g., Barnosky et al., 2012; Steffen et al., 2018). Therefore, a rudimentary cost-benefit analysis argues strongly for caution and strong actions to mitigate global warming and species extinction.

Climate Change Scenarios

A key challenge for the prognosis is the limited knowledge of climate sensitivity. Climate sensitivity is the increase in the global mean equilibrium temperature as a result of an increase in climate forcing. It is independent of the nature of the change in the forcing and gives the increase in temperature for a unit change in forcing. In connection with increases in atmospheric CO_2 it is often given as the temperature increase for a doubling of atmospheric CO_2. The range of climate sensitivity is between 1.5 and 4.5°C and the most likely value is between 2.5 and 3.0°C (Pachauri et al., 2014, and the references therein).

The large range in climate sensitivity introduces a significant uncertainty in models used to compute trajectories of possible future climates. Another uncertainty is in the future anthropogenic emissions of greenhouse gases. To address this uncertainty, ensemble studies utilizing a large number of Earth system climate models in combination with a large set of emission scenarios have been carried out (e.g., Pachauri et al., 2014). The results indicate a wide range of temperature

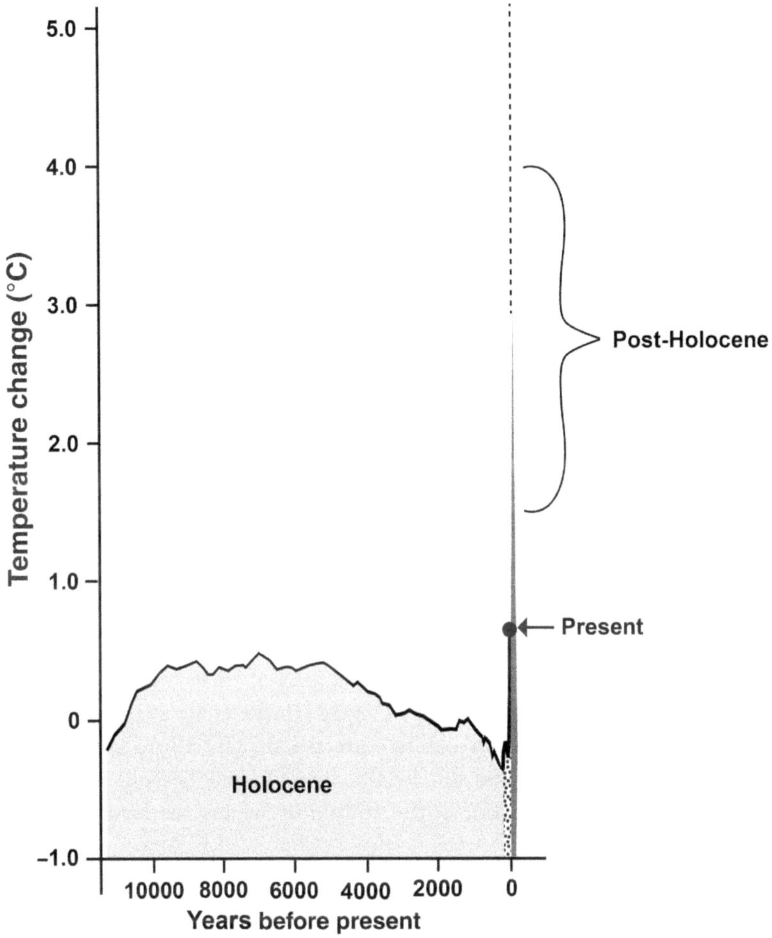

FIGURE 1.2 Possible temperature changes in the 21st century compared with Holocene temperature variations. The total range of global mean temperature variation during the Holocene up to 1900 was on the order of 1°C. The likely change of 2°C or more by 2100 would cause a rapid transition into a post-Holocene unknown to humanity. The dotted area indicates the time from 1880 to 2000.

Source: The temperature curve for the Holocene is based on Figure 1b in Marcott et al. (2013).

increases by 2100 with large regional variability. However, the likely increase will put the Earth's life-support system in a state unknown to humanity (see Figure 1.2 and Pachauri et al., 2014; IPCC, 2018).

The changes in the dynamics of the climate system triggered by the higher energy level have the potential to lead to changes in the regional distribution of floods and droughts. Model studies (e.g., Paltan et al., 2018) and recent observed changes (e.g., Wang et al., 2018) demonstrate the potential for large changes in regional patterns of precipitation, which can lead to extreme floods in areas that have not experienced such floods before, and to extreme droughts where droughts have been virtually unknown before. It appears likely that the magnified shift in temperatures will bring more concurrent temperature and drought extremes in the future, exacerbating individual impacts from heat waves and droughts (Chiang et al., 2018).

A global existential threat not addressed by the scenario studies mentioned above is associated with the passing of possible climate tipping points, which, if these points exist, could lead to a very different climate state. Steffen et al. (2018) point out that climate does not change linearly but rather in steps, which potentially could be large and catapult the Earth system into a hothouse state very detrimental to the human population. As a result, the single-species high-energy pulse comes with the potential to push the climate system over a tipping point separating the current climate from a hothouse state.

Impacts on Humanity

Climate change impacts on humans are not equally distributed across the planet, and it is very likely that the impacts will increase inequality. For example, the impacts of heat waves on workers are expected to be much larger in the developing world than in the developed world (Burke et al., 2015). There is also injustice in the sense that many of the countries that have least contributed to the cause of modern global and climate change will likely be burdened with the most severe consequences, while at the same time having the least resources to cope with them.

"Climate change is fundamentally redrawing the map of where people can live" (Solheim & Swing, 2018). Mass migration is already posing a global-scale problem and this challenge is very likely going to increase rapidly. Migration will also result from a potentially significant reduction in global carrying capacity, which could cause large-scale famine (Spratt & Dunlop, 2018).

A major economic risk is associated with sea-level rise. In many areas, population has moved into the coastal zone, and most of the megacities are located in coastal areas with little topography. Even a modest increase in sea level could result by 2100 in annual flood costs on the order of tens of trillion US dollars (Jevrejeva et al., 2018). However, an economic crisis of unparalleled scale could emerge much earlier if risk perception concerning sea-level rise changes and the currently high value of coastal real estate property evaporates.

Increased storminess can amplify the extensive disturbances that storms cause to marine ecosystems and habitats that support productive fisheries (Sainsbury et al., 2018). The impact of changing storms on fisheries will vary spatially due to spatial variations in projected changes in storm risk, changes in target fish species, the resilience of infrastructure and the extent of natural and man-made storm defenses. It is expected that the impacts will be larger for small-scale fisheries (Sainsbury et al., 2018). Considering that 3.1 billion people get close to 20% of their animal protein from fish, and fish are relied on for vital micronutrients, a negative impact of changes in storminess on fisheries can have a significant impact on food security.

The expected increase in atmospheric CO_2 is predicted to have an impact on the nutritional value of major food sources. Many food crops grown under atmospheres with CO_2 contents of 550 ppm have significantly reduced protein, iron, and zinc contents compared with current conditions (Smith & Myers, 2018). The likely increase in atmospheric CO_2 over the next decades could lead to an additional more than 100 million people being protein deficient, and many more impacted by a significant reduction of dietary iron, particularly in high risk regions such as South and Southeast Asia, Africa, and the Middle East (Smith & Myers, 2018).

The most devastating impact of modern climate change on humanity could result from a significant reduction of the Earth's carrying capacity. Lovelock (2009) estimated that by 2050 the global carrying capacity may be down to 1 billion humans. The fossil fuel-based industrial agriculture created a transient increase in carrying capacity in terms of food and allowed in the last century the addition of 5.5 billion individuals to the already existing population of 1.5 billion humans. It must be doubted that this increase in carrying capacity can be maintained indefinitely. In fact, it is not clear that the carrying capacity was actually increased, since it resulted from a large-scale use of non-renewable resources (including land and soil) and caused widespread soil degradation. Providing water and food to sustain the growing population is increasingly challenging (de Marsily & Abarca-del Rio, 2016). Moreover, the rapidly growing population is exceeding the carrying capacity in almost all other aspects including land-use (Barnosky et al., 2012), mineral resources (Assadourian, 2013), and harvesting the biosphere (Smil, 2011).

From an analytical outside look at an ecosystem typically applied by ecologists, it is easily recognized if a population exceeds the carrying capacity of this system. However, from a more experimental inside look strictly limited to the human population, it might appear that humans have not exceeded the carrying capacity because there are still enough individuals to produce offspring and increase the population size. Importantly, the point where the carrying capacity is exceeded is long before the impacts are felt and a complete collapse of the population is unavoidable (Figure 1.3). A transient increase in carrying capacity may delay the collapse, but it also increases the overshoot and the size of the subsequent catastrophic population collapse significantly. For humanity, the point of crossing the unmodified planetary carrying capacity appears to be some time before industrial agriculture facilitated a large transient increase. Modern global and climate change may make it increasingly difficult to sustain this transient increase in the carrying capacity to further delay the collapse.

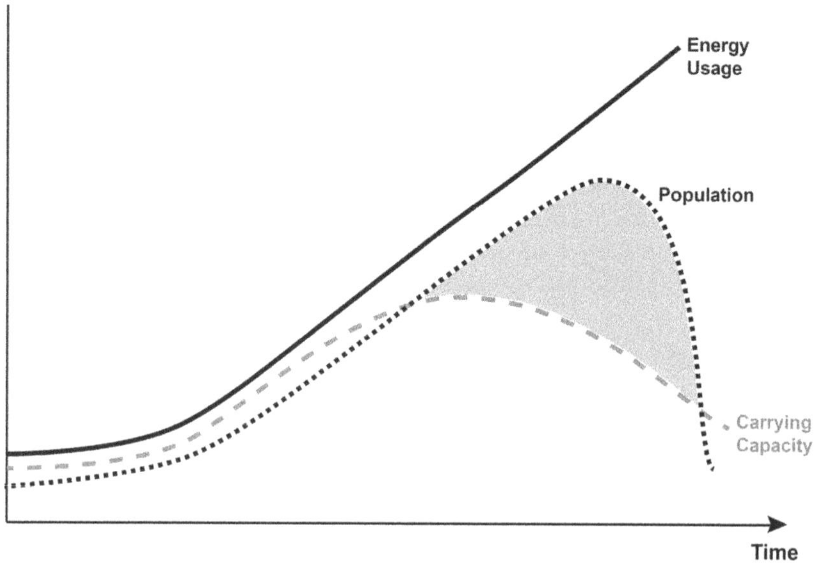

FIGURE 1.3 Impact of a transient increase in Earth's carrying capacity on population. Humanity has used its easy access to seemingly unlimited energy to facilitate a transient increase in Earth's carrying capacity, which sustained a rapid population growth most likely beyond the carrying capacity (gray area). The likely reduction of the carrying capacity will increase the overshoot and the severity of the population collapse.

Impacts on the Earth's Life-Support System

Ecosystems react to climate change by altering structure, behavior, and movements. For example, poleward extension of the tropics is a likely consequence of a warming climate (Staten et al., 2018). The extent of anthropogenic land use changes combined with climate change has the potential to initiate a state shift of the global biosphere (Barnosky et al., 2012). When species cannot tolerate climate change *in situ*, or migrate and colonize suitable habitat elsewhere quickly enough, they become extinct (Nogués-Bravo et al., 2018). The ability of species to adapt depends on the speed of changes. The option of migration also depends on the availability of suitable migration paths and locations. Anthropogenic climate disruption is predicted to soon compete with habitat destruction as the most important driver of contemporary extinctions (Nogués-Bravo et al., 2018). Without substantial mitigation efforts, terrestrial ecosystems are at risk of major transformation in composition and structure. Records of terrestrial vegetation since the last glacial period indicate that terrestrial ecosystems are highly sensitive to temperature change (Nolan et al., 2018). Considering that the projected temperature changes are much faster than those during the warming after the last ice age, it is highly likely that modern climate change puts terrestrial ecosystems worldwide at risk of major transformation, impacting biodiversity and disrupting ecosystem services (Nolan et al., 2018).

The anticipated changes in many aspects of climate could also lead to possible formation of novel climates with conditions that have no current or past analog (Williams & Jackson, 2007). This prospect, which entails that existing habitats would disappear and new reconfigured ones would arise, puts limits on the capability to describe the transitions the Earth's life-support system might be heading to and to develop foresight covering all possible futures (Kittel, 2013). Scenario-based studies consistently indicate that biodiversity will continue to decline throughout the 21st century, but the range of projected changes is still very broad because of major policy opportunities for interventions and large uncertainties in projections (Pereira et al., 2010).

Therapy

Addressing the Climate Symptom

Reduction of greenhouse gas emissions is widely considered to be the most important step for mitigating climate change, and a wide range of countries and organizations are focusing on a decarbonization of human systems. The commitments made as part of the Paris Agreement indicate the growing understanding of the importance of this action. Adopted in 2015, the Paris Agreement set the specific goal of keeping global warming well below 2°C above pre-industrial levels, and of pursuing efforts to limit the warming to 1.5°C. However, the most recent report published by the Intergovernmental Panel on Climate Change (IPCC, 2018) concludes that:

> Pathways limiting global warming to 1.5°C with no or limited overshoot would require rapid and far-reaching transitions in energy, land, urban and infrastructure (including transport and buildings), and industrial systems (high confidence). These systems transitions are unprecedented in terms of scale, but not necessarily in terms of speed, and imply deep emissions reductions in all sectors, a wide portfolio of mitigation options and a significant upscaling of investments in those options (medium confidence).

Considering the urgency of the required transitions in the next ten years, it is highly unlikely that humanity will be willing and able to thoroughly reorganize basically everything.

A number of geoengineering approaches have been proposed to reduce the incoming solar radiation reaching the Earth's surface, to accelerate CO_2 absorption in major sinks, to extract CO_2 from the atmosphere, or to increase ice accumulation in the Arctic to limit sea level rise, among others. Most of these proposals have potentially severe side effects and could easily lead to failure with severe consequences afterwards (e.g., Cotton-Barratt et al., 2016). A recent proposal to cover a very large part of the Sahara desert with solar panels both to utilize the electrical power generated and to change the surface albedo to increase rainfall and thus reverse the expansion of the desert caused by climate change (Li et al., 2018) is among the more realistic proposals. A promising removal strategy would utilize biogeochemical improvement of soils by adding crushed, fast-reacting silicate rocks to croplands (Beerling et al., 2018). This approach could lead to improved crop production, increased protection from pests and diseases, and restoration of soil fertility and structure. However, none of the climate geoengineering proposals appears to have the potential to sufficiently reduce emissions (Lawrence et al., 2018).

A major transition of building material from concrete and steel to wood also provides an opportunity to reduce carbon emission. A significant transition to a meat-reduced diet on global scale could further reduce greenhouse gas emissions in the form of CO_2 and CH_4. However, Rieder (2016) concludes that reducing procreation would be the single most efficient step to reduce greenhouse gas emissions.

Aiming for Sustainability

The global community is pursuing a quest for sustainable development, which in 2015 resulted in the agreement on seventeen Sustainable Development Goals in the 2030 Agenda for Sustainable Development (United Nations, 2015). Only one of these goals focuses on mitigation of climate change and adaptation to an emerging new climate, while several others are related to climate change drivers. Likewise, very few of the goals aim at reducing the flows in the Earth's life-support system. In the process leading to the Agenda, several proposals were made for goals that acknowledged the importance of safeguarding the Earth's life-support system and keeping it within the boundaries of the safe operating space (e.g., Griggs et al., 2013), but these proposals were not agreeable to the intergovernmental United Nations.

A number of fundamental transformations have been identified that would facilitate progress towards the Sustainable Development Goals (e.g., Utting, 2016; TWI2050 – The World in 2050, 2018). However, all of these transformations are anthropocentric and lack a deep understanding of humanity as an integral part of the Earth's life-support system together with all other species. While these transformations might slow down the rapid degradation of the health of the planetary life-support system and to some extent slow down climate change, they are by far not sufficient to reverse the degradation and improve the system's health.

Importantly, none of these transformations addresses the fact that the current economic model discounts the future and is focused on immediate needs and benefits. The growth-dependent mainstream economic model is a major obstacle for progress towards more sustainability and mitigating global and climate change. Climate change policies often are focused on increasing energy efficiency. However, efforts to decouple economic growth from impacts on the life-support system by increasing efficiency as part of a "green economy" are likely to fail due to the Jevons paradox (e.g., Jevons, 1866; Freire-González & Puig-Ventosa, 2015). This is a rebound effect in which the reduction of costs in energy (or resource) services actually leads to higher consumption as a result of increased efficiency. Model-based studies show that the rebound effect leads to an increase of resource usage far beyond the Earth's carrying capacity (Font Vivanco et al., 2016). Consequently, within an economic system that favors and demands growth, the cataclysmic energy pulse created by *Homo sapiens* can hardly be ended or reduced by an increase of efficiency, and the world as a whole is not on a track towards "greener growth" or a "green economy" (Giljum et al., 2014).

Ending the Energy Pulse

The most promising avenue to thoroughly limiting the impact of the single-species high-energy pulse on the Earth's life-support system is implicit in the definition of sustainable development as provided by Griggs et al. (2013). Recognizing the role of the mainstream economic model in the regulation of flows between human communities and the Earth's life-support system, the purpose of economy needs to be extended from the creation of human wealth to the safeguarding of the Earth's life-support system (Plag & Jules-Plag, 2017). With this extension, economic activities would inherently aim to meet the needs of human communities while at the same time limit the flows between these communities and the planetary life-support system and ensure that the flows are not degenerating the health of this system. Applying ethical principles to assess the morality of human interactions with the life-support system, as done, for example, by Rieder (2016), would provide additional justifications for this fundamental change in the purpose and goals of economy. This would transform the anthropogenic virus into the healer of the life-support system. By elevating the safeguarding of the planetary life-support system to being the purpose of economy, it seems possible for humanity to change its function from a potentially catastrophic virus in the Earth's life-support system to the "healer" of this system. In fact, without this transformation of the purpose of economy, it appears likely that a major state shift of the planetary life-support system will lead to a socio-economic collapse with a deconstruction of the current governance structures that ensure some level of civilization (Bendell, 2018).

Concluding Remarks

The extraordinary brains of *Homo sapiens* have enabled this species to outcompete all other mammals and to re-engineer the planetary life-support system to an extent very few other species were able to achieve in Earth's history. And those other highly successful species often modified the system to an extent that made it no longer suited for their own survival, particularly after thresholds were crossed. On the other hand, species that have low metabolic rates and are least impactful on their environment seem to have persisted over much longer times. The basal metabolic rate "could therefore represent an important metric for predicting future extinction patterns, with changes in global climate potentially affecting the lifespan of individuals, ultimately leading to the extinction of the species they are contained within" (Strotz et al., 2018). This raises the question of whether the principles of evolution actually limit the chances of a single exceptionally successful species to take control of the Earth's life-support system without pushing the system outside the safe operating space for this species. The extraordinary success of *Homo sapiens* in developing and maintaining a virtual metabolic rate far above the rate within the scaling law for mammals created a cataclysmic energy pulse that modified the Earth's life-support system fundamentally. The planetary system is being pushed into a new high-energy climate state, a hothouse state (Steffen et al., 2018), unsuitable for most mammals to survive. This success may be the basis for the demise of this—and many other—species.

As hard as it may be to accept this, modern climate change could turn out to be a process that ends the current high-energy pulse and the global dominance of a single species, thus restoring some inter-species equity and justice in the Earth's life-support system. Garnett (2018) asks whether the modern human impact on all aspects of the Earth's life-support system is leading to a total systemic failure of this system and suggests that humanity currently lacks "the tools and analytical capacity to understand the significance of these changes and therefore" cannot answer this question. This incapacity of fully assessing the human impact on the Earth's life-support system should lead humanity to a very careful progress and should put focus on "less being better," instead of trying to facilitate more economic growth and to sustain a rapidly increasing global population. Reducing the global population through a change in the morality of procreation, as concluded by Rieder (2016), would be at the core of mitigating modern global and climate change and safeguarding the Earth's life-support system.

References

Arrhenius, S., 1896. On the influence of carbonic acid in the air upon the temperature of the ground, *Philosophical Magazine and Journal of Science*, **41**, 237–276.

Assadourian, E., 2013. Re-engineering cultures to create a sustainable civilization, in *State of the World 2013: Is Sustainability Still Possible?*, edited by the Worldwatch Institute, Chapter 10, Island Press, Washington, DC.

Barnosky, A. D., Hadly, E. A., Bascompte, J., Berlow, E. L., Brown, J. H., Fortelius, M., Getz, W. M., Harte, J., Hastings, A., Marquet, P. A., Martinez, N. D., Mooers, A., Roopnarine, P., Vermeij, G., Williams, J. W., Gillespie, R., Kitzes, J., Marshall, C., Matzke, N., Mindell, D. P., Revilla, E., & Smith, A. B., 2012. Approaching a state shift in earth's biosphere, *Nature*, **486**, 52–58.

Barnosky, A. D., Brown, J. H., Daily, G. C., Dirzo, R., Ehrlich, A. H., Ehrlich, P. R., Eronen, J. T., Fortelius, M., Hadly, E. A., Leopold, E. B., Mooney, H. A., Peterson Myers, J., Naylor, R. L., Palumbi, S., Stenseth, N. C., & Wake, M. H., 2014. Introducing the scientific consensus on maintaining humanity's life support systems in the 21st century: Information for policy makers, *Anthropocene Review*, **1**(1), 78–109.

Bedsworth, L., Cayan, D., Franco, G., Fisher, L., & Ziaja, S., 2018. Statewide summary report. California's fourth climate change assessment, Publication number: Sum-ccca4-2018-013, California Governor's Office of Planning and Research, Scripps Institution of Oceanography, California Energy Commission, California Public Utilities Commission.

Beerling, D. J., Leake, J. R., Long, S. P., Scholes, J. D., Ton, J., Nelson, P. N., Bird, M., Kantzas, E., Taylor, L. L., Sarkar, B., Kelland, M., DeLucia, E., Kantola, I., Müller, C., Rau, G., & Hansen, J., 2018. Farming with crops and rocks to address global climate, food and soil security, *Nature Plants*, **4**(3), 138–147.

Bendell, J., 2018. *Deep adaptation: A map for navigating climate tragedy*, IFLAS Occasional Paper 2, www.iflas.info, July 27, 2018.

Bianchi, T. S. & Morrison, E., 2018. Human activities create corridors of change in aquatic zones, *Eos*, **99**. DOI: 10.1029/2018EO104743.

Bishop, M. J., Mayer-Pinto, M., Airoli, L., Firth, L. B., Morris, R. L., Loke, L. H. L., Hawkins, S. J., Naylor, L. A., Coleman, R. A., Chee, S. Y., & Dafforn, K. A., 2017. Effects of ocean sprawl on ecological connectivity: Impacts and solutions, *Journal of Experimental Marine Biology and Ecology*, **492**, 7–30.

Bojinski, S., Verstrate, M., Peterson, C. P., Richter, C., Simmons, P., & Zemp, M., 2014. The concept of essential climate variables in support of climate research, applications, and policy, *Bulletin of the American Meteorological Society*, **95**(9), 1431–1443.

Bond, D. P. G. & Grasby, S. E., 2017. On the causes of mass extinctions, *Palaeogeography, Palaeoclimatology, Palaeoecology*, **478**, 3–29.

Breitburg, D., Levin, L. A., Oschlies, A., Grégoire, M., Chavez, F. P., Conley, D. J., Garcon, V., Gilbert, D., Gutiérrez, D., Isensee, K., Jacinto, G. S., Limburg, K. E., Montes, I., Naqvi, S. W. A., Pitcher, G. C., Rabalais, N. N., Roman, M. R., Rose, K. A., Seibel, B. A., Telszewski, M., Yasuhara, M., & Zhang, J., 2018. Declining oxygen in the global ocean and coastal waters, *Science*, **359**(6371), 7240.

Bryden, H. L., Longworth, H. R., & Cunningham, S. A., 2005. Slowing of the Atlantic meridional overturning circulation at 25°N, *Nature*, **438**, 655.

Burke, M., Hsiang, S. M., & Miguel, E., 2015. Global non-linear effect of temperature on economic production, *Nature*, **527**, 235–239.

Caesar, L., Rahmstorf, S., Robinson, A., Feulner, G., & Saba, V., 2018. Observed fingerprint of a weakening Atlantic Ocean overturning circulation, *Nature*, **556**, 191–196.

Capra, F., 1996. *The Web of Life*, Random House Publishers.

Ceballos, G., Ehrlich, P. R., Barnosky, A. D., García, A., Pringle, R. M., & Palmer, T. M., 2015. Accelerated modern human-induced species losses: Entering the sixth mass extinction, *Science Advances*, **1**(e1400253), 1–5.

Cheng, L., Abraham, J., Hausfather, Z., & Trenberth, K. E., 2019. How fast are the oceans warming? *Science*, **363**(6423), 128–129.

Chiang, F., Mazdiyasni, O., & AghaKouchak, A., 2018. Amplified warming of droughts in southern United States in observations and model simulations, *Science Advances*, **4**(8).

Costanza, R., Alperovitz, G., Daly, H., Farley, J., Franco, C., Jackson, T., Kubiszewski, I., Schor, J., & Victor, P., 2013. Building a sustainable and desirable economy-in-society-in-nature, in *State of the World 2013: Is Sustainability Still Possible?* edited by the Worldwatch Institute, Chapter 11, Island Press, Washington, DC.

Cotton-Barratt, O., Farquhar, S., Halstead, J., Schubert, S., & Snyder-Beattie, A., 2016. *Global catastrophic risks 2016*, Tech. rep., Global Challenge Foundation and Global Priorities Project.

de Marsily, G. & Abarca-del Rio, R., 2016. Water and food in the twenty-first century, *Surveys in Geophysics*, **37**(2), 503–527.

Drake, T. W., Wickland, K. P., Spencer, R. G. M., McKnight, D. M., & Striegl, R. G., 2015. Ancient low-molecular-weight organic acids in permafrost fuel rapid carbon dioxide production upon thaw, *Proceedings of the National Academy of Sciences* **112**, 13946–13951.

EASAC, 2013. *Extreme weather events in Europe: preparing for climate change adaptation*, EASAC report number 22, European Academies Science Advisory Council.

EASAC, 2018. *Extreme weather events in Europe: preparing for climate change adaptation: an update on EASAC's 2013 study*, Tech. rep., European Academies Science Advisory Council.

Font Vivanco, D., Kemp, R., & van der Voet, E., 2016. How to deal with the rebound effect? A policy-oriented approach, *Energy Policy*, **94**, 114–125.

Frade, P. R., Bongaerts, P., Englebert, N., Rogers, A., Gonzalez-Rivero, M., & Hoegh-Guldberg, O., 2018. Deep reefs of the Great Barrier Reef offer limited thermal refuge during mass coral bleaching, *Nature Communications*, **9**(1), 3447.

Freire-González, F. & Puig-Ventosa, I., 2015. Energy efficiency policies and the Jevons Paradox, *International Journal of Energy Economics and Policy, Econjournals*, **5**(1), 69–79.

Gaffney, O. & Steffen, W., 2017. The Anthropocene equation, *The Anthropocene Review*, **4**(1), 53–61, DOI: 10.1177/2053019616688022.

Garnett, P., 2018. Total systemic failure? *Science of the Total Environment*, **626**, 684–688.

Giljum, S., Dittrich, M., Lieber, M., & Lutter, S., 2014. Global patterns of material flows and their socio-economic and environmental implications: An MFA study on all countries world-wide from 1980 to 2009, *Resources*, **3**, 319–339.

Griggs, D., Stafford-Smith, M., Gaffney, O., Rockström, J., Öhman, M. C., Shyamsundar, P., Steffen, W., Glaser, G., Kanie, N., & Noble, I., 2013. Sustainable Development Goals for people and planet, *Nature*, **495**, 305–307.

Haddad, N. M., Brudvig, L. A., Clobert, J., Davies, K. F., Gonzalez, A., Holt, R. D., Lovejoy, T. E., Sexton, J. O., Austin, M. P., Collins, C. D., Cook, W. M., Damschen, E. I., Ewers, R. M., Foster, B. L., Jenkins, C. N., King, A. J., Laurance, W. F., Levey, D. J., Margules, C. R., Melbourne, B. A., Nicholls, A. O., Orrock, J. L., Song, D.-X., & Townshend, J. R., 2015. Habitat fragmentation and its lasting impact on earth's ecosystems, *Science Advances*, **1**(2).

Hansen, J., Sato, M., Kharecha, P., Beerling, D., Masson-Delmotte, V., Pagani, M., Raymo, M., Royer, D. L., & Zachos, J. C., 2008. Target atmospheric CO2: Where should humanity aim? *The Open Atmospheric Science Journal*, **2**, 217–231.

Hansen, J., Sato, M., Kharecha, P., & von Schuckmann, K., 2011. Earth's energy imbalance and implications, *Atmospheric Chemistry and Physics*, **11**, 13421–13449.

Harbitz, C., Løvholt, F., & Bungum, H., 2014. Submarine landslide tsunamis: how extreme and how likely? *Natural Hazards*, **72**(3), 1341–1374.

Herculano-Houzel, S., 2017. The remarkable (but not extraordinary) human brain, *Scientific American Mind*, **28**(1), 36–41, https://www.scientificamerican.com/article/lessons-from-making-brain-soup/.

Higman, B., Shugar, D. H., Stark, C. P., Ekström, G., Koppes, M. N., Lynett, P., Dufresne, A., Haeussler, P. J., Geertsema, M., Gulick, S., Mattox, A., Venditti, J. G., Walton, M. A. L., McCall, N., Mckittrick, E., MacInnes, B., Bilderback, E. L., Tang, H., Willis, M. J., Richmond, B., Reece, R. S., Larsen, C., Olson, B., Capra, J., Ayca, A., Bloom, C., Williams, H., Bonno, D., Weiss, R., Keen, A., Skanavis, V., & Loso, M., 2018. The 2015 landslide and tsunami in Taan Fiord, Alaska, *Scientific Reports*, **8**(1), 12993.

Hughes, T. P., Kerry, J. T., Baird, A. H., Connolly, S. R., Dietzel, A., Eakin, C. M., Heron, S. F., Hoey, A. S., Hoogenboom, M. O., Liu, G., McWilliam, M. J., Pears, R. J., Pratchett, M. S., Skirving, W. J., Stella, J. S., & Torda, G., 2018. Global warming transforms coral reef assemblages, *Nature*, **556**, 492–496.

IPCC, 2018. *Global Warming of 1.5°C, Special Report 15*, Intergovernmental Panel on Climate Change. https://www.ipcc.ch/sr15.

Jevons, W. S., 1866. *The coal question: an inquiry concerning the progress of the national and the probable exhaustion of our coal-mines*, Macmillan and Company, London, 2nd edn.

Jevrejeva, S., Jackson, L. P., Grinsted, A., Lincke, D., & Marzeion, B., 2018. Flood damage costs under the sea level rise with warming of 1.5°C and 2°C, *Environmental Research Letters*, **13**(7).

Kittel, T. G. F., 2013. The vulnerability of biodiversity to rapid climate change, in *Vulnerability of Ecosystems to Climate*, edited by R. A. Pielke, Sr., T. R. Seastedt, & K. Suding, vol. 4 of *Climate Vulnerability: Understanding and Addressing Threats to Essential Resources*, pp. 185–201, Elsevier Inc., Academic Press, Oxford.

Kohnert, K., Serafimovich, A., Metzger, S., Hartmann, J., & Sachs, T., 2017. Strong geologic methane emissions from discontinuous terrestrial permafrost in the Mackenzie Delta, Canada, *Scientific Reports*, **7**(1), 2045–2322.

Laffoley, D. and Baxter, J. M. (eds.), 2016. *Explaining ocean warming: Causes, scale, effects and consequences,* Tech. rep., International Union for Conservation of Nature and Natural Resources, Available at https://portals.iucn.org/library/node/46254.

Lawrence, M. G., Schäfer, S., Muri, H., Scott, V., Oschlies, A., Vaughan, N. E., Boucher, O., Schmidt, H., Haywood, J., & Scheffran, J., 2018. Evaluating climate geoengineering proposals in the context of the Paris Agreement temperature goals, *Nature Communications*, **9**(1), 3734. DOI: 10.1038/s41467-018-05938-3.

Li, Y., Kalnay, E., Motesharrei, S., Rivas, J., Kucharski, F., Kirk-Davidoff, D., Bach, E., & Zeng, N., 2018. Climate model shows large-scale wind and solar farms in the Sahara increase rain and vegetation, *Science*, **361**(6406), 1019–1022.

Lovelock, J., 2009. *The Vanishing Face of Gaia*, Basic Books, New York.

Marcott, S. A., Shakun, J. D., Clark, P. U., & Mix, A. C., 2013. A reconstruction of regional and global temperature for the past 11,300 years, *Science*, **339**, 1198–1201, DOI: 10.1126/science.1228026.

Molena, F., 1912. Remarkable weather of 1911 – the effect of combustion of coal on the climate – what scientists predict for the future, *Popular Mechanics*, March 1912, 393–342.

Morrison, A. E., 2017. State of the climate in 2016, *Bulletin of the American Meteorological Society*, **98**(8), Si–S280.

National Research Council, 2013. *Abrupt Impacts of Climate Change: Anticipating Surprises*, Washington, D.C., The National Academies Press.

Nogués-Bravo, D., Rodríguez-Sánchez, F., Orsini, L., de Boer, E., & Jackson, S. T., 2018. Cracking the code of biodiversity responses to past climate change, *Trends in Ecology and Evolution* **33**(10), 765–776.

Nolan, C., Overpeck, J. T., Allen, J. R. M., Anderson, P. M., Betancourt, J. L., Binney, H. A., Brewer, S., Bush, M. B., Chase, B. M., Cheddadi, R., Djamali, M., Dodson, J., Edwards, M. E., Gosling, W. D., Haberle, S., Hotchkiss, S. C., Huntley, B., Ivory, S. J., Kershaw, A. P., Kim, S.-H., Latorre, C., Leydet, M., Lézine, A.-M., Liu, K.-B., Liu, Y., Lozhkin, A. V., McGlone, M. S., Marchant, R. A., Momohara, A., Moreno, P. I., Müller, S., Otto-Bliesner, B. L., Shen, C., Stevenson, J., Takahara, H., Tarasov, P. E., Tipton, J., Vincens, A., Weng, C., Xu, Q., Zheng, Z., & Jackson, S. T., 2018. Past and future global transformation of terrestrial ecosystems under climate change, *Science*, **361**(6405), 920–923.

Pachauri, R. K., Allen, M. R., & Barros, V. R. et al., 2014. *Climate Change 2014: Synthesis Report. Contribution of Working Groups I, II, and III to the Fifth Assessment Report of the Intergovernmental Panel on Climate Change (IPCC)*, IPCC.

Paltan, H., Allen, M., Haustein, K., Fuldauer, L., & Dadson, S., 2018. Global implications of 1.5°C and 2°C warmer worlds on extreme river flows, *Environmental Research Letters*, **13**(9), 094003. DOI: 10.1088/1748-9326/aad985.

Pearce, F., 2010. From ocean to ozone: Earth's nine life-support systems, *New Scientist*, Feature, 24 February 2010.

Pereira, H. M., Leadley, P. W., Proença, V., Alkemade, R., Scharlemann, J. P. W., Fernandez-Manjarrés, J. F., Araújo, M. B., Balvanera, P., Biggs, R., Cheung, W. W. L., Chini, L., Cooper, H. D., Gilman, E. L., Guénette, S., Hurtt, G. C., Huntington, H. P., Mace, G. M., Oberdorff, T., Revenga, C., Rodrigues, P., Scholes, R. J., Sumaila, U. R., & Walpole, M., 2010. Scenarios for global biodiversity in the 21st century, *Science*, **330**(6010), 1496–1501.

Pimm, S. L., Jenkins, C. N., Abell, R., Brooks, T. M., Gittleman, J. L., Joppa, L. N., Raven, P. H., Roberts, C. M., & Sextonet, J. O., 2014. The biodiversity of species and their rates of extinction, distribution, and protection, *Science*, 344.

Plag, H.-P. & Jules-Plag, S.-A., 2017. An economy for humanity: Transition to an economy for a thriving humanity and planetary future, *ApoGeoSpatial*, **32**(2, Spring 2017), 30–35.

Reed, K. A., Stansfield, A. M., Wehner, M. F., & Zarzycki, C. M., 2018. *The human influence on Hurricane Florence*, Tech. rep., Stony Brook University, School of Marine and Atmospheric Sciences.

Rieder, T. N., 2016. *Toward a Small Family Ethics – How Overpopulation and Climate Change Are Affecting the Morality of Procreation*, Springer.

Rind, D., Lean, J. L., Lerner, J., Lonergan, P., & Leboissetier, A., 2008. Exploring the stratospheric/tropospheric response to solar forcing, *Journal of Geophysical Research*, **113**, D24103.

Ripple, W. J., Wolf, C., Newsome, T. M., Galetti, M., Alamgir, M., Crist, E., Mahmoud, M. I., Laurance, W. F., & 15,364 scientist signatories from 184 countries, 2017. World scientists warning to humanity: A second notice, *BioScience*, **67**(12), 1026–1028.

Rockström, J., Steffen, W., Noone, K., Persson, Å., Chapin, F. S. I., Lambin, E., Lenton, T. M., Scheffer, M., Folke, C., Schellnhuber, H., Nykvist, B., De Wit, C. A., Hughes, T., van der Leeuw, S., Rodhe, H., Sörlin, S., Snyder, P. K., Costanza, R., Svedin, U., Falkenmark, M., Karlberg, L., Corell, R. W., Fabry, V. J., Hansen, J., Walker, B., Liverman, D., Richardson, K., Crutzen, P., & Foley, J., 2009. A safe operating space for humanity, *Nature*, **461**, 472–475.

Rothman, D. H., 2017. Thresholds of catastrophe in the earth system, *Science Advances*, **3**(9), e1700906.

Ruddiman, W. F., 2005. How did humans first alter global climate? *Scientific American*, **292**(3), 46–53.

Sainsbury, N. C., Genner, M. J., Saville, G. R., Pinnegar, J. K., O'Neill, C. K., Simpson, S. D., & Turner, R. A., 2018. Changing storminess and global capture fisheries nature climate change, *Nature Climate Change*, **8**(8), 655–659.

Smil, V., 2011. Harvesting the biosphere: The human impact, *Population and Development Review*, **37**(4), 613–636.

Smith, A., 1776. *An Inquiry into the Nature and Causes of the Wealth of Nations. Volume 1*, W. Strahan, London.

Smith, M. R. & Myers, S. S., 2018. Impact of anthropogenic CO2 emissions on global human nutrition, *Nature Climate Change*, **8**(9), 834–839.

Solheim, E. & Swing, W. L., 2018. Making migration safe for climate nexus, *Project Syndicate – The World's Opinion Page*, Aug 21, 2018.
Spratt, D. & Dunlop, I., 2018. *What lies beneath. The understatement of existential climate risk*, Breakthrough National Centre for Climate Restoration, Melbourne.
Starr, M., 2018. Our planet's largest mass extinction had warning signs – and they're happening again, *Science Alert*, March 29, 2018.
Staten, P. W., Lu, J., Grise, K. M., Davis, S. M., & Birner, T., 2018. Re-examining tropical expansion, *Nature Climate Change*, **8**(9), 768–775.
Steffen, W., Richardson, K., Rockström, J., Cornell, S. E., Fetzer, I., Bennett, E. M., Biggs, R., Carpenter, S. R., de Vries, W., de Wit, C. A., Folke, C., Gerten, D., Heinke, J., Mace, G. M., Persson, L. M., Ramanathan, V., Reyers, B., & Sörlin, S., 2015. Planetary boundaries: Guiding human development on a changing planet, *Science*, **347**(6223).
Steffen, W., Rockström, J., Richardson, K., Lenton, T. M., Folke, C., Liverman, D., Summerhayes, C. P., Barnosky, A. D., Cornell, S. E., Crucifix, M., Donges, J. F., Fetzer, I., Lade, S. J., Scheffer, M., Winkelmann, R., & Schellnhuber, H. J., 2018. Trajectories of the earth system in the Anthropocene, *Proceedings of the National Academy of Sciences*, **115**(33), 8252-8259.
Strotz, L. C., Saupe, E. E., Kimmig, J., & Lieberman, B. S., 2018. Metabolic rates, climate and macroevolution: a case study using Neogene molluscs, *Proceedings of the Royal Society of London B: Biological Sciences*, **285**(1885).
Thornalley, D. J. R., Oppo, D. W., Ortega, P., Robson, J. I., Brierley, C. M., Davis, R., Hall, I. R., Moffa-Sanchez, P., Rose, N. L., Spooner, P. T., Yashayaev, I., & Keigwin, L. D., 2018. Anomalously weak Labrador Sea convection and Atlantic overturning during the past 150 years, *Nature*, **556**, 227–230.
Timmermans, M.-L., Toole, J., & Krishfield, R., 2018. Warming of the interior Arctic Ocean linked to sea ice losses at the basin margins, *Science Advances*, **4**(8).
Trenberth, K. E., Fasullo, J. T., & Balmaseda, M. A., 2014. Earth's energy imbalance, *Journal of Climate*, **27**(9), 3129–3144.
TWI2050 – The World in 2050, 2018. *Transformations to Achieve the Sustainable Development Goals*, Tech. rep. International Institute for Applied Systems Analysis (IIASA), Laxenburg, Austria. www.twi2050.org.
UNEP, 2017. *The Emissions Gap Report 2017*, Tech. rep., United Nations Environment Programme (UNEP), Nairobi.
Union of Concerned Scientists, 1992. *World scientists warning to humanity*, Tech. rep., Union of Concerned Scientists USA.
United Nations, 2015. *Transforming our world: The 2030 agenda for sustainable development*, Tech. Rep. A/RES/70/1, United Nations.
Utting, P., 2016. Chapter 4, promoting social and solidarity economy through public policy, in *UNRISD Flagship Report 2016: Policy Innovations for Transformative Change – Implementing the 2030 Agenda for Sustainable Development*, edited by K. Hujo, pp. 115–140, UNRISD, Geneva.
von Schuckmann, K., Palmer, M. D., Trenberth, K. E., Cazenave, A., Chambers, D., Champollion, N., Hansen, J., Josey, S. A., Loeb, N., Mathieu, P.-P., Meyssignac, B., & Wild, M., 2016. An imperative to monitor earth's energy imbalance, *Nature Climate Change*, **6**, 138–144.
Wallström, M., Bolin, B., Crutzen, P., Steffen, W., & International Herald Tribune, 2004. A global crisis: The Earth's life-support system is in peril, *The New York Times*, 20 January.

Wang, X.-Y., Li, X., Zhu, J., & Tanajura, C. A. S., 2018. The strengthening of Amazonian precipitation during the wet season driven by tropical sea surface temperature forcing, *Environmental Research Letters*, **13**(9), 094015. DOI: 10.1088/1748-9326/aadbb9.

West, G., Brown, J. H., & Enquist, B. J., 1997. A general model for the origin of allometric scaling laws in biology, *Science*, **276**(5309), 122–126.

Widmaier, E. P., Raff, H., & Strang, K. T., 2016. *Vander's Human Physiology: The Mechanisms of Body Function*, McGraw-Hill Education, New York, NY.

Williams, J. W. & Jackson, S. T., 2007. Novel climates, no-analog communities, and ecological surprises, *Frontiers in Ecology and the Environment*, **5**(9), 475–482.

Wuebbles, D.J., Fahey, D.W., Hibbard, K.A. Dokken, D.J. Stewart, B.C. and May-Cock, T.K. (eds.), 2017. *Climate science special report: Fourth national climate assessment*, Volume I, U.S. Global Change Research Program, Washington, DC, USA.

Young, O. & Steffen, W., 2009. The earth system: Sustaining planetary life-support systems, in *Principles of Ecosystem Stewardship*, edited by C. Folke, G. Kofinas, & F. Chapin, Springer, New York, NY.

2

ETHICAL CHALLENGES POSED BY CLIMATE CHANGE

An Overview

Madison Powers

Overarching Issues

Climate change poses well-known threats to human life, health, and habitats. Over the long term, global atmospheric warming threatens the planetary systems that have made possible all life on Earth over the last 10,000 years of the Holocene era (Steffen et al. 2015; Powers 2018). Even in the near term, the projected adverse effects include species extinction; drought, desertification and disruption of the hydrologic cycle; sea level rise; extreme weather events and disrupted growing seasons; crop loss from flooding and fresh water runoff, especially where agriculture is dependent upon rainfall; and expansion and worsening of geographic zones at highest risk for infectious diseases (IPCC 2014). These adverse effects are already being felt by, and will be most consequential for, the nations that are the poorest, hottest, agriculturally most vulnerable to weather pattern disruption, economically most dependent on agriculture, most vulnerable to vector borne diseases that are expected to increase dramatically, and least able to adapt by virtue of their disadvantageous geography and lack of economic resources.

It is not surprising, then, that many of the pressing moral issues pertain to the assignment of responsibility for addressing the current and expected effects of anthropogenic climate change. The complex cluster of environmental, economic, and political problems emerging from the accumulation and concentration of greenhouse gases in the Earth's atmosphere challenges the capacity of traditional moral theories to provide practical guidance. Particular strands of a complex puzzle are often best illuminated within the context of specific theoretical approaches to moral problem solving, each of which brings some aspect of the problems to the forefront. The framework of subsequent sections of this chapter examines various moral theories sequentially in order to isolate specific issues in the assignment of moral responsibility.

However, the processes leading to climate change raise more general moral questions. Key issues include identification of the appropriate target for harm prevention, the distinction between duties of mitigation and duties of adaptation, recognition that the duties will conflict or efforts to fulfill them will be self-defeating or working at cross-purposes, and decisions regarding the distribution of the costs and other economic burdens of fulfilling those duties.

The Target of Harm Prevention

The first issue is the appropriate target for coordinated international efforts to limit the likely harm to the planet overall and specific regions. It is often said that climate change poses a special, if not unique moral challenge because "dangerous anthropogenic climate change" only materializes once a threshold level of greenhouse gas concentration is reached (UNFCCC 1992). After years of discussion about the appropriate target for limiting the concentration of accumulated greenhouse gases, the political consensus emerging in 2015 from the 21st Conference of Parties in Paris, endorsed the goal of restricting atmospheric greenhouse gas concentration to 450 parts per million (ppm) (UNFCCC 2015). This figure corresponds roughly to 1 trillion metric tons of greenhouse gases in the atmosphere, the point at which the Earth's atmospheric temperature would be raised by 2 degrees Celsius above pre-industrial levels. In part, this choice of target reflects a judgment of acceptable risk, predicated on estimates of a threshold of significant danger of irreversible environmental impacts threatening the stability of the planet.

However, harms to human health and environment from climate change are occurring now. After considerable protest from representatives of low-lying and island nations, the text of the Agreement acknowledged that considerable harm will occur at a 1.5 degree increase in atmospheric temperature. Moreover, recent research suggests that at the current 400 ppm level of concentration considerable harm is resulting from sea level rise and weather pattern disruption, thereby increasing frequency and magnitude of storm damage and diminishing agricultural production (Schurer et al. 2017).

The upshot is that the debate over the targets of harm prevention reflects a deep normative disagreement over what constitutes acceptable harm. Representatives of regions of the world who are hurt first and worst by climate change want the world to commit to doing more, sooner, while other parties to the negotiations favor a less aggressive response, reflecting a reluctance to undertake expensive and socially disruptive behavioral change.

Duties of Mitigation and Adaptation

The second issue pertains to alternative perspectives on the kinds of moral responses that are most pressing. On the one hand, an emphasis on mitigation of harm is prudent not only for the sake of limiting the risk of irreversible changes in the envelope of stability of the life-supporting planetary system, but for the sake of those experiencing harm at current levels. The task of harm mitigation involves slowing of the rate of increase in the stock of atmospheric greenhouse gases, and ultimately, reversing globally risky and regionally destructive levels of concentration.

Mitigation can be achieved primarily in either of two ways. The analogy to a bathtub is often used to make the point. We can close the valve at the tap or accelerate the flow through the drain. Concretely, the analogy highlights two distinct mitigation strategies. We can lower the atmospheric concentration below the current 400 ppm level either by reducing the rate of greenhouse gas emissions or enhancing mechanisms of greenhouse gas absorption in forests or oceans so that less ends up in the atmosphere where it causes harm.

Mitigation has long-term benefits for everyone, and it offers hope for near-term relief for countries already bearing some of the burdens of climate change. However, mitigation has an immediate adverse effect on affluent countries. Successful mitigation strategies will likely require a massive shift away from forms of production and consumption based on fossil fuels. Absent technological breakthroughs and widespread diffusion of cheaper and cleaner energy sources, the likely side effect is a reduction in standard of living.

Mitigation also is achievable through absorption – paradigmatically natural absorption through forest preservation – but at the moment, it comes with high costs and uncertain benefits for countries that are asked to forgo economic growth and expansion through natural resource development.

Mitigation also might be achieved by some form of geoengineering. The goal is the extraction of greenhouse gases from the atmosphere or redirection of energy into oceans or beyond the atmosphere. Critics view such strategies with skepticism, not only due to doubts regarding technological feasibility and scalability, but on the basis of moral concerns about the wisdom of conducting an unprecedented experiment with potentially catastrophic consequences. These worries are heightened by geoengineering's reliance on speculative theoretical assumptions and the lack of procedural safeguards and mechanisms for political accountability (Gardiner 2011, 339–396).

An alternative to costly and socially disruptive mitigation duties places greater emphasis on duties to facilitate adaptation to new heat and rainfall patterns, the increase in extreme weather events, and rising sea levels. Those hurt first and worst would have to alter existing agricultural practices, build more resilient buildings and infrastructure, or move away from coastlines and river beds. Because many countries lack the economic resources or technological capacities to undertake the necessary steps to adapt, the discharge of such duties inevitably must fall to richer countries or others with greater resources. The discharge of such duties might be undertaken through transfers of financial resources or reduced cost or no-cost technology transfers that enable those who are harmed to absorb or adapt to the environmental shocks. In short, neither mitigation nor adaptation is likely without considerable economic burdens for those best positioned to bear them.

Competing Strategies

Given that some harms are baked into the current levels of atmospheric greenhouse gases, the need for adaptation to changed environments is unavoidable. Absent a more robust commitment to more timely fulfilment of duties of mitigation, even more attention to adaptation strategies will be necessary to alleviate the predicted increase in human suffering. Adaptation strategies, however, often place decision makers in a bind. Three examples are illustrative.

Use of existing technologies for adaptation often cuts against the goal of greenhouse gas emissions reduction. As countries get hotter, especially in middle-income countries like India, greater reliance on refrigeration and indoor air conditioning will mean more emissions for the sake of adjusting to heat-related threats to agricultural products in the countryside and human life in the cities (Isaac and van Vuuren 2009). Adaptation strategies therefore sometimes run counter to the goals of mitigation.

Moreover, adaptation strategies sometimes compete with one another. In India, for example, climate change is exacerbating loss of both available surface water (e.g., rivers and streams) and ground water (i.e., aquifer) reserves. The response to the failure of surface water sources to replenish is to drill deep bore wells in search of water reserves in the aquifers. However, over-pumping has led to considerable reduction in recharge rates of aquifers (*The Economist* 2009). Strategies for adaptation to current loss of accessible water thus undermine the availability of water reserves that will be needed to adapt to long-term adverse hydrologic changes associated, in part, with climate change.

Finally, there will be instances in which adaptation duties cannot be feasibly and fully satisfied through resource and technology transfers from resource-rich to resource-poor nations. Some areas of the world will become radically inhospitable to human settlements, resulting in new waves of global migration. Estimates range from 200 million to 1 billion by 2050, with more recent predictions topping 1.4 billion by 2060 (Geisler and Currens 2017). Whatever the number turns out to be, in some cases, adaptation duties can only be satisfied by opening national borders for access to vital, life-sustaining resources, including water and habitable and arable land. The fulfilment of adaptation duties by the addition of new residents increases domestic energy demands, thereby adding to the challenge of meeting their own mitigation goals.

Bearing the Burdens of Mitigation and Adaptation

It is clear that effective and timely mitigation efforts and successful adaptation strategies will likely require citizens of affluent states to provide poor countries with the resources that they lack, or make sacrifices in their standard of living. In one sense, it seems fair to place the burden on rich countries that are home to some of the world's biggest emitters. Indeed, many affluent nations became affluent and remain affluent by burning fossil fuels. Current generations can hardly complain about being asked to make some sacrifice of the enormous benefits that they have inherited. However, there is an imperfect correlation between resource-rich countries and high-emissions countries. The share of contributions to the existing stocks made by the world's richest industrialized countries over the span of the last 250 years is on schedule to be overtaken by the recent and ongoing contributions of rapidly industrializing middle-income countries (OECD 2008). Many of the high-emissions countries best positioned to undertake emissions reductions on the scale needed to avert serious harms from climate change are neither the historical beneficiaries of fossil fuel wealth nor emitting currently for the sake of maintaining a high standard of living. Their high emissions are for the sake of lifting historically unprecedented numbers of people out of poverty and ensuring better life prospects for future generations.

More generally, harms produced by climate change do not fit neatly into some of our standard ways of understanding moral responsibility. If the same consequences were the predictable side effects of a high-tech military conflict conducted in space by superpowers, most would agree that the warring parties have a stringent moral duty to refrain from such actions, however much it disadvantages them, and to compensate innocent parties because of the harm inflicted on innocent bystanders. If the same consequences were produced by an asteroid or tsunami, many would argue for the existence of stringent duties of beneficence to people who suffer losses from environmental disaster, based primarily on ability to bear the costs.

Nonetheless, those most affected are likely to experience the initial brunt of the harms of climate change very differently from a comparable loss due to an asteroid or tsunami. They are likely to feel a sense of injustice. They will know that the source of the harms is anthropogenic. While they are not in precisely the same moral posture as innocent bystanders, harmed by the actions of warring nations, most of the individuals hurt first and worst reside in countries that have contributed least to the problem. All the while, many others located elsewhere have benefitted disproportionately and continue to benefit from the very activities that result in the misfortune of modest contributors. The sense of injustice is likely to be magnified by the knowledge that what everyone does or fails to do damages the life prospects for the entire planet, that there is nothing that these countries can do on their own to prevent these outcomes, and that many of the hardest hit are least able to bear the burdens. They are effectively trapped in an unfolding spiral of harm that can only be averted by sacrifice by those who benefit most (Powers 2015).

Political rhetoric notwithstanding, the reality is that we are not "all in it together." Favorably circumstanced parties would have to agree to give up quite a lot of clear benefits now in order to avert a massive and imminent harm to other parties, with only the expectation of a less certain benefit in the distant future. Put another way, the motivational basis is lacking for a mutually beneficial agreement based on a game-theoretic strategy of tit-for-tat reciprocity – a decision context in which the parties are symmetrically situated, where both parties believe they are giving and receiving comparable value in return.

Assignment of Moral Responsibility

The main normative challenge, then, is to identify a plausible rationale for the assignment of responsibility for mitigating harm and adaptation to the harms that mitigation efforts fail to stop. In moral theory generally, the path toward an answer typically begins with a presumption that causal contributors stand first in line for bearing responsibilities of both sorts. Similarly, within environmental ethics and discussions of the normative foundations of environmental tort law, the basic idea is often described as the Polluter Pays Principle (PPP). Such a principle is appealing for a simple reason. It answers to the intuitive idea that all and only the causal contributors to some problem should have primary duties to mitigate ongoing harm, prevent future harm, and remedy or compensate for the harms created by their actions (Adler 2007; Perry 1992; Caney 2010a).

Laddered Approaches

Ordinarily, we look first to agents that have some moral culpability, typically based on a wrongful intention or a failure to exercise a duty of due care or a duty to prevent harm to others.

While nation-states and other agents who stand in some special relationship of responsibility for the well-being of vulnerable parties might have back-up duties, the proximate causal agents are first in line. When individual agents are unable or not in a position of legitimate moral authority to take the necessary steps to prevent or remedy particularly grave or urgent harms, a long philosophical tradition recognizes a variety of arguments for a moral duty to establish institutions with the requisite organizational capacities and normative authority to tackle problems that assignment of responsibility to causally implicated individuals cannot resolve (Kant 1996; Pufendorf 1994; Rawls 1999, 99). The analysis of the proper locus of moral responsibility thus begins with the individual but ascends to another level, where assignment of responsibility is pushed upward to an institutional agent or other back-up agent.

The examination of some unusually complex questions of moral responsibility, including climate change, global poverty, and other harms produced within networks of interacting causal chains, involving multiple contributing agents, proceeds by way of this laddered approach. If the first step up the ladder proves inadequate to the resolution of a moral problem, then the focus of inquiry ascends to the next level, typically an institutional agent, such as a nation-state, primarily because it is where the requisite institutional capacities and normative authority for coordinated action reside (Meckled-Garcia 2008).

Although the moral rationales for the laddered approach differ somewhat, the oldest and most familiar versions of the argument focus on duties to create states. However, the argument is readily expanded to accommodate the prospect that state institutions also might prove inadequate. We assign responsibility by ascending to the next level, for example, by creating new institutional arrangements beyond the nation-state (Wenar 2007; Ronzoni 2009; Scheffler 2010). The rationale undergirding the laddered approach has been used to argue for taking yet another step up the ladder, calling for a global constitutional convention to address competing intergenerational claims presented by climate change, on the assumption that internationally negotiated treaties, as currently envisioned, are inadequate to the task (Gardiner 2014). The central line of argument therefore mirrors more familiar versions of the laddered approach.

In each instance, the locus of responsibility shifts upward, often to new institutional agents that are better suited to the tasks of deliberation and coordination and equipped with the relevant expertise and authority to take the kinds of action otherwise likely to lack effectiveness or moral legitimacy.

Moral Pessimism

The search for a theoretical rationale for the assignment of responsibility for climate change harm proceeds against the backdrop of considerable doubt about the prospects for success at any step on the ladder. Some moral philosophers argue that climate change is a complex moral problem which our ordinary accounts of individual and institutional responsibility are ill-equipped to handle (Jamieson 2010, 2014; Sinnott-Armstrong 2010). Dale Jamieson, for example, asserts that our inherited conception of individual moral responsibility "presupposes that harms and their causes are individual, that they can be readily identified, and that they are local in time and space" (Jamieson 2010, 83). However, dangerous anthropogenic global warming is a function of the aggregate effects of uncoordinated actions undertaken by billions of individuals, firms, and governments, often for benign or salutary ends, over the course of centuries. The harm resulting from climate change therefore is the consequence of many small causal contributions made in the process of generating and using electricity, building cities, driving cars, cutting trees, and so on (Jamieson 2010; Sinnott-Armstrong 2010). No single contribution is the necessary or sufficient cause of the harm any identifiable individuals will experience. Moreover, because many of the emissions produced today will dissipate over 200–300 years we cannot assume that they will contribute to the problem in the distant future.

The conclusion of these pessimistic arguments is that whatever happens, however bad, might turn out to be a human tragedy, for which literally no one can be said to be culpable.

Three Generic Challenges

Those who think traditional moral theories face potentially insurmountable challenges for the assignment of moral responsibility for both individual and institutional agents, under any moral theory, often point to a cluster of problematic features of standard conceptions of moral responsibility. The problems are thought to occur at every step up the ladder, pushing us toward further ascent, only to face roughly the same challenges, as well as new ones at each new rung. Three of these generic arguments pose important challenges, but none seems decisive.

Determining When Harming is Wrong

The first of these challenges proceeds from the fact that there are many salutary uses of greenhouse gas emissions. Accordingly, it is implausible to think that greenhouse gas emission is in itself wrong. It is wrong, if at all, only contingently within a context in which the joint product of multiple agents, generally acting without coordination, produces a harm or unacceptable risk of harm (Sinnott-Armstrong 2010). However, the fact that a moral theory has to come to grips with the thorny problem of specifying the conditions under which some harm or imposition of an elevated risk of harm is wrong is neither an unusual problem nor necessarily insurmountable.

J.S. Mill, for example, famously argued that the mere fact that a business enterprise draws customers away from "disappointed competitors" does not entail a wrong or an injustice (Mill 1977, 293). Some setback to the interests of others is an inevitable feature of organized social life. Therefore, we have to rely upon other morally salient considerations, such as reasonable expectations or prior claims of right, to determine whether a harm should be considered a wrong. Even ill intention is not always decisive. I might open my pizza parlor across the street from a popular franchise of a national chain. I might intend to put it out of business by producing superior pizzas, and I might even intend to expand my business to multiple competing locations, and one by one, watch the dominos fall. Profits, jobs, and livelihoods will be lost.

The same point can be made about the imposition of risk in cases where actual harm might not occur. Economists point to the pervasive existence of negative externalities – costs and risks imposed on third parties by every kind of activity, including ones that have countervailing, socially beneficial purposes. We have no choice – legally or morally – but to decide which "harms to notice" (Baier 1985). We typically do so either by some process of balancing expected harms and benefits or by defining specific domains of permissible private action, in which the costs or risks imposed upon others are not treated as relevant grounds justifying interference with an activity.

In sum, all moral theories face the challenge of explaining which harms caused in the process of morally benign or socially beneficial activities count as wrongs.

The existence of socially beneficial emissions that contribute to the harms from dangerous anthropogenic climate change is a similar problem to those already faced by moral philosophers and legal theorists (Wenar 2007).

Fair Apportionment

The second generic challenge contends that the basis for apportionment of causal responsibility among so many causal contributors, over an extended and ongoing time frame, is so indefinite and speculative that it is neither feasible nor fair to single out some contributors (Caney, 2010b, 207; Posner and Sunstein 2008, 18). In particular, the worry is that the complete identification of the members of a set of putative wrongdoers is thwarted by the fact that some individuals are no longer living and some corporate entities are no longer doing business (Posner and Sunstein 2008, 18; Caney, 2010a, 130; Miller 2009, 126–27).

In essence, these critics are arguing that fairness requires holding responsible all or none. Some of these arguments in the context of moral theory turn on an analogy to legal theories of compensation for injury. For example, consider cases in which several manufacturing companies dump chemical effluent into a river over a period of several years, and it is clear that the parties do not act in a coordinated manner, or even act within the same time frame. In addition, over time some of the effluent degrades or is absorbed into the ground and dispersed and diluted in larger bodies of water. However, the cumulative consequence of a certain kind of conduct by many separate agents is a toxic soup that damages farmland downstream. Some companies are no longer in business, some companies profited greatly, and other companies barely made ends meet. Some companies dumped a lot of chemical effluent, and others much less.

However, in recent years the trend has been to dismiss the "all or none" objection. Many legal jurisdictions concede that it is unfair to assign liability without clear criteria for apportioning damages on the basis of contribution, but the conclusion is that it would be a graver form of unfairness to leave injured parties without recourse to remedy (Perry 1992). A more egregious form of unfairness would arise from letting everyone off the hook, simply because it is not possible to hold all contributors accountable or accountable in proportion to their contribution. More generally, this conclusion is not a departure from ways we already think of fairness norms. It is often recognized within moral theory that, when not all forms of unfairness can be avoided, competing claims of unfairness sometimes have to be ranked in comparative terms, rather than treated as preemptive considerations (Broome 1990).

The Question of Culpability

A third challenge points to a lack of a plausible account of moral culpability in the emitters' actions, especially for emissions in the distant past. Arguably, very few causal contributors to climate change, at any stage of history, are culpable either because of malign intent or because their actions were inherently unjust in the way slavery can be said to be wrong in itself (Sinnott-Armstrong 2010; Miller 2009, 129). Even the weaker notion of negligence in tort law, from which we might construct a moral analogue, is thought by some to be too problematic for holding current generations responsible for past greenhouse gas emissions that remain in the atmosphere. As a condition for imposing liability, the argument is that it requires a showing of harm from some conduct that was in violation of existing norms of due care, which the parties being held responsible either knew or should have known they were violating at the time of emission (Posner and Sunstein 2008, 18–19).

For the sake of argument, suppose that we lack proper basis for holding emitters accountable for emissions prior to some recent date, when it became reasonable to conclude that emitters knew or should have known about the consequences of greenhouse gas accumulation. At some point, it is reasonable to assume that greenhouse gas emissions, for some purposes, on some scale, and in light of available feasible alternatives to achieve salutary purposes, constitutes a sufficiently anti-social activity, for which exculpatory considerations begin to fade from moral relevance.

All three generic arguments in support of the conclusion that existing moral theories suffer from intractable problems of assigning responsibility for greenhouse gas emissions fail. Determining when harming or imposing risk of harm constitutes a wrong is not a problem different in kind from ones routinely faced by moral theories. Any unfairness in holding some but not all causal contributors responsible for an aggregate harm might be outweighed or overridden by the importance of avoiding an even graver form of unfairness. Culpability need not be predicated upon malicious intent, and we have ample theoretical resources from which to develop alternative grounds for culpability. Such a judgment would have to be based on a complex evaluation of scale of emissions, purposes of emissions, and feasible alternatives for emitters who we can safely assume to be aware or who should be aware of the consequences of their actions.

Harming, Wronging, and Doing Wrong

Lingering in the background are deeper challenges that moral theories face in dealing with climate change issues, based on the differences in the way each type of theory conceptualizes the nexus between harming someone, wronging someone, and doing wrong.

Harming and Moral Wrong

One reason for doubt about the prospects of existing moral theories to deal with issues of climate change responsibility focuses on the absence of a tight linkage between a moral wrong and causing harm to someone. The challenge is embodied in what Derek Parfit calls the "person-affecting principle." The intuitive idea is that "bad" acts must be "bad for" someone (Parfit 1984, 363). In order to assess whether someone has been harmed, Parfit says that we should employ a historical counterfactual test. A wrong action (or policy) is one that harms someone in the sense that it makes someone worse off. Parfit's person-affecting principle is meant to be ecumenical among various theories of what makes someone better off or worse off. The counterfactual test might employ notions of well-being such as pleasure, happiness, or a pluralist conception.

If we concur with Parfit's claim that a tight link between harming and moral wrong has a strong intuitive grounding, we encounter two major obstacles in making sense of the wrongness of certain kinds of activities.

The first problem arises in cases of the sort Sinnott-Armstrong highlights. He considers the emissions from a recreational Sunday drive in an automobile. "No storms or droughts or heat waves can be traced to my individual act of driving" (Sinnott-Armstrong 2010, 336). Even if we are drawn to the thought that there is something morally problematic about fun but frivolous joy rides in gas guzzling automobiles in the current environmental context, it is not clear that any identifiable person is made worse off by that action, and thus unclear what grounds we have for moral criticism.

The second problem is that even if for the sake of argument, my activities, perhaps together with thousands or millions of other Sunday drivers, do reduce the quality of life for generations in the distant future, we still lack a basis for concluding that anyone was made worse off in the way that the counterfactual test of the person-affecting principle presupposes. Parfit and others refer to this worry as the non-identity problem.

Consider what we can say about the massive greenhouse gas emissions of all of the leisurely driving in 2020 and the quality of life of persons born several generations later. At the time of the action, there is literally *no one* whose quality of life is made lower by those emissions. Even if the quality of life of future generations is quite low, and that outcome is a direct causal consequence of our actions today, there are no persons who can say that they have been harmed by our actions or energy use policies. This is because everything we do today determines the identity and number – and even whether – people are born in the distant future. If we accept the person-affecting principle, we cannot say that anyone has been wronged, simply because there is no one who has been made worse off.

There seem to be two main options for getting around the counterintuitive implications. We might retain the core of the person-affecting principle by jettisoning the historical counterfactual test of harming. Instead of saying that persons are harmed when made worse off than they would have been but for a causally intervening act, the relevant notion of harm might be replaced, or more plausibly, supplemented by an alternative conception. We might say that persons are harmed by being made worse off than they are entitled to be, for example, as assessed by reference to a minimum quality of life standard set by a theory of human rights (Pogge 2007). We are thereby able to hold on to the tight link between harming and moral wrong, but we can do so only by revising the notion of what counts as a harming. The problem with this strategy is that it is unclear what the revised notion of harm adds to the understanding of wronging. Pogge's point can be made more straightforwardly. All he has to claim is that wronging can be a matter of violating some independent moral standard.

The thought that wronging need not involve harming leads to a second option for getting around the strictures of the person-affecting principle. Moral wrong might be defined disjunctively. We might think of moral wrongs as either involving harm in the way the person-affecting principle supposes, or involving some other kind of wrong-making feature of the relationship between wrongdoers and those who are wronged. Because we already have on hand some familiar moral theories that reject the exclusivity of the harm-based account of moral wrong embodied in the person-affecting principle, it is worth exploring how some of them might theorize the relation between activities that contribute to dangerous climate change and wronging others.

Wronging Someone without Harming

The tradition of moral theorizing rooted in Kant's moral theory offers a number of options for how we might unpack his conception of wronging persons. For example, we wrong others by failing to treat them with the respect due to a dignified being, treating them as mere means, or failing to recognize them as self-directing agents.

One concrete suggestion along these lines is found in Scanlon's contractualist theory (Scanlon 1998). His claim is that we should understand moral wrong as a matter of violating a norm that no one could reasonably reject as the basis for interpersonal relations. If this is a plausible way of understanding what makes wrong acts wrong, there is no need for the idea of harming of a person in order to understand the wrongness of an action. Critics nonetheless will complain that more needs to be said in order to elucidate what constitutes reasonable rejectability in general, and what sort of actions that contribute to adverse environmental impact fit the bill. However, if we can fill out the story in a satisfactory way, then we can speak meaningfully of wronging others, including future generations, without having to pass the counterfactual test of identifying anyone in particular who has been made worse off in the way Parfit's person-affecting principle supposes (Kumar 2003).

Another alternative to the reduction of all wronging to harming others involves the invasion of the sovereign domain of decision making or violating their reasonable expectations of exclusive control (Ripstein 2006). In a similar vein, the cumulative actions of high-emissions individuals and countries could be seen as involving a unilateral imposition of risk to the most vital interests of the global poor who will be hurt first and worst. A morally salient fact is their position of utter powerlessness to secure their own future, and to some extent, we can understand their situation as one in which they are in the grip of others in ways that resemble paradigmatic forms of unjust domination (Laborde 2010).

Among other tasks, such approaches would have to say more about the matters that fall within someone's sovereign domain of decision making with regard to climate impacts, and how the aggregated negative externalities of the discrete actions of so many other agents constitute unjust forms of control. Moreover, powerlessness is a condition in which one's fate is out of one's control, but being powerless is not equivalent to the unjust relations picked out by theories of domination. Such theories focus on the wrongness of an agent's having arbitrary and unilateral power over a subordinate party. Application of standard theories of domination would have to accommodate the fact that the powerlessness felt by relatively modest contributors to the cumulative stock of greenhouse gases is not solely due to the power wrongly exercised by other agents. Even comparatively modest emitters are implicated in their own inability to forestall harm to themselves and to everyone else. There is, in short, no sharp divide between victims and wrongdoers that allows us to track the distinction between dominant and subordinate parties.

Perhaps a third line of Kantian-inspired argument might bypass the challenge posed by the lack of a standard for wrongful greenhouse gas production and the fact that even the actions of the least environmentally destructive individuals and communities contribute to the condition of helplessness for themselves and others. Various substantive notions of failures of moral reciprocity suggest a more concrete way of elucidating the general Kantian idea of a failure to respect the agency of others and treat them in ways appropriate to their standing as moral equals. Ranier Forst, for example, observes that a minimum commitment to reciprocity at the heart of an ideal of moral equality requires "that none of the parties concerned may claim certain rights and privileges it denies to others" (Forst 2001, 177). In the context of climate change, ideals of reciprocity might recognize the existence of some permissible level of contribution to the stock of greenhouse gases, such that the low-end users among the global poor are not properly counted as participating in a moral wrong. High-end users above that level, by contrast, would be wrongdoers were they to insist on their own prerogative to emit for the sake of a higher standard of living, and if pursuit of that higher standard deprives the poor of the most basic requirements of life that the affluent would similarly claim for themselves. If a strategy of this sort proves successful, then we have both insight into who is wronged and the standard by which wrongness might be assessed.

Doing Wrong without Harming or Wronging Someone

Thus far, reciprocity theories, as well as reasonable rejectability theories, and domination or invasion of moral sovereignty theories provide only placeholder accounts for how climate change duties might be grounded. A critic might well demand more in the way of practical guidance. Still lacking are substantive answers to questions about the quantity of emissions and their purposes that allow us to judge that someone is wronged.

However, Parfit argues that the theoretical path forward lies elsewhere (Parfit 1984, 384). He says we need an impersonal way to explain what makes a wrong act wrong. An impersonal theory is one that views wrong acts as wrong for reasons other than someone's having been made worse off or wronged. If successful, such a strategy would articulate a standard of conduct, below which we can be said to do wrong. If successful, the point made against Pogge earlier is that we need not worry about whether there is any specific person who is wronged. We would simply have a theory of what constitutes wrongdoing without having to assume the need to identify who is harmed or wronged. The lack of a link between doing wrong and an identifiable wronged person – a harm to their well-being interests, a violation of sovereign rights, being disrespected or disadvantaged, being subject to unjust dominion, a failure of reciprocity, or whatever – does not matter because the test of right and wrong is a matter of compliance with or violation of a general norm.

To be sure, it matters to each person who can say that because of some action or policy she has been wronged. She has a special standing to complain, but the wrong she experiences is not a function of the fact that she in particular has been wronged. Tim Hayward presses what I take to be a similar point. He asks what moral difference it makes that the duty breached is a duty directed "to me." It is the departure from a standard of due care that is the defining feature of the wrong, not the identity of the persons wronged (Hayward 2013). No fact peculiar to specific identifiable right-holders – contemporaneous "statistical others" or future generations – provides grounds for such rights. The grounds are "general facts about what is good for humans" and the kinds of conduct that exhibit a strong tendency to undermine their most vital interests (Hayward 2013, 280). Rahul Kumar makes a similar point. All that is necessary to know is what it is that persons in specific situations are entitled to expect of other persons in showing respect for the moral status of persons (Kumar 2003, 111).

Participation in Collective Harming

An alternative approach attributes moral responsibility to individuals on the basis of their participation in a collective activity rather than any feature that explains the wrongness of particular actions. The standard model of complicity in collective injustice supposes more than a mere joint product; it requires a joint project. Paradigmatically, that involves two things: participants work together, each doing some part, knowing that others are doing their part toward the achievement of a shared end, and the shared end itself is morally condemnable.

Christopher Kutz suggests that both assumptions should be relaxed (Kutz 2000). For example, civilian railway conductors who coordinated the train schedules delivering people to Nazi death camps should not be let off the hook either by claiming that their participation made no difference to the outcome or that they only intended to collect a paycheck, and did not share the Nazi's corrupt intention.

Assuming both revisions are plausible, the ascription of moral responsibility for participation in collective wrongdoing in standard cases like the railway conductor case is brought closer to cases that Kutz calls unstructured collective harms. His example of an unstructured collective harm involves numerous car drivers whose actions have a cumulative harmful effect – the environmental harm of damage to the Earth's protective ozone layer. Kutz's revisionist construal of collective wrongdoing thus resembles instances in which individuals participate in social processes that produce climate change. Participants in the joint production of these environmental harms are in a similar moral position as the railway conductor. They share no culpable intention and make little or no causal difference to the outcome. On Kutz's account, we should not hold the drivers blameless for their participation in the creation of a collective harm, for much the same reason we should not hold similarly situated Nazi-era railway workers blameless.

The problem with Kutz's argument is that cases of unstructured collective environmental harms are importantly different from the case of the railway conductor under the Nazi regime. We still need an explanation of what makes participation culpable, and in standard cases, participants are implicated by their complicity with the actions of one or more agents who bear clear marks of moral wrongdoing. Presumably, there are no agential analogues to the Nazis in either the climate change or ozone depletion cases. The existence of at least one morally corrupt agent in the network of contributing agents is what casts a moral shadow over the character of those who unintentionally but knowingly aid the evil project of others. Kutz, however, argues that the locus of what is wrong-making in this scenario is morally deficient character, not a morally culpable discrete action, shared intention, joint project, or causal difference. However, for that strategy to work, we have to identify what remains an elusive basis for our negative assessment of character, given that our character is not impugned by our contributions to the evil projects of others.

Green Virtues

Others who doubt the prospects for identifying what makes actions contributing to the harms caused by climate change wrong also suggest shifting to considerations of character under a virtue ethics approach. Ronald Sandler, for example, argues that instead of looking at individual actions and their harmful consequences we should be looking at character traits that point to the "green" virtues. Sandler observes that a "character trait is a virtue to the extent that its possession is generally conducive to promoting the good; and a character trait is a vice to the extent that it is generally detrimental to promoting the good" (Sandler 2010, 176). The underlying rationale for identifying virtue is similar to Julia Driver's theory, which defines virtue as a character trait that "systematically produces or gives rise to the good" (Driver 2001, 108).

Dale Jamieson recommends this approach to those who are inclined toward utilitarian moral theories. Instead of looking at the consequences of discrete actions, or even the aggregated consequences of multiple actions, we can assess the character traits of people whose actions over a lifetime reveal dispositions that we have reason to condemn, and accordingly, grounds for assigning them some greater share of responsibility relative to those whose dispositions tend to be environmentally less destructive (Jamieson 2007). The obvious merit of the virtue approach is that it does not depend on the existence of a joint project, a wrongful intention, a shared end, or a causally efficacious action resulting in a harm to an identifiable victim. Vice is simply a character flaw because of its systematic tendency to produce grave harm.

Green virtues, however, are not easy to identify, even with the aid of an appeal to broadly consequentialist underpinnings. Virtue is usually construed as a dispositional mean between two vices. Courage, for example, is the mean between cowardice and recklessness. Benevolence is a mean between the overly generous dispositions of a spendthrift and the miserly dispositions of persons lacking in compassion for others. But how do we mark the "green mean"? Intuitively, we might be inclined to identify some clear patterns of excess, at least relative to some well-defined purpose. We might think that setting the thermostat at maximum performance levels when no one is at home is evidence of culpable environmental indifference. Beyond these rather simple, context- and purpose-dependent activities, it is by no means clear what would count, say, as general virtues with respect to the production of greenhouse gases.

Perhaps in the light of the urgency of climate change threats to the planet, the virtuous person would aim for no net greenhouse gas production. We might reconstruct John Broome's suggestion that we each have duties to zero out our carbon footprint because anything else amounts to an injustice (Broome 2012). Not only are there empirical disagreements about whether specific carbon offset programs do what they promise, but existing energy technology makes such offsets feasible for no more than a small sliver of the Earth's population. As a result, we would run into the traditional worry about the plausibility of virtues that depend too much on luck or good fortune. Currently, it simply is not possible for everyone, or perhaps even very many, to be virtuous in this manner.

Alternatively, some notion of virtuous emitters might be gleaned from a reflection on Henry Shue's distinction between subsistence and luxury emissions (Shue 2010). Shue's aim is to answer the question of how much emissions exceed an individual's entitlement to a declining common pool resource. He argues that, given that we are well beyond 60% of the 1 trillion metric tons of accumulated atmospheric greenhouse gas concentration consistent with a livable, 2-degree target, current emission patterns of the global affluent are not justifiable (Shue 2011). We might reconstruct Shue's argument to say that using more than needed to sustain life itself is evidence of a serious character flaw. However, not all individual usage is for one's own benefit or for superfluous consumption aspirations. For example, some carbon emissions are for the sake of ameliorating life-threatening poverty or many other salutary purposes.

Driver suggests we should be searching for a "character trait [that] is a virtue to the extent that its possession is generally conducive to promoting the good." In that case it is not clear that the focus on green virtues, in isolation from consideration of the totality of benefits and harms from all of our consumption and production activities, will offer helpful practical guidance, either for how individuals should live or for identifying culpable participation in an unstructured collective harm.

Therefore, we need a more detailed explanation of what distinguishes virtuous and condemnable participation in a vast and complex causal network of agents who differ in intention, social utility of their activities, and so on. For a notion of virtue to fill that gap, some independent standard of excess and unjustified greenhouse gas contribution is needed in order to assess character.

Nation-States and International Institutions

There is an important sense in which the collectively produced impacts of climate change are more structured than the foregoing discussions suggest. Individual and corporate greenhouse gas footprints are what they are only because of the state's fingerprints in establishing the legal and regulatory background. Moreover, states are uniquely positioned to alter the conduct of many separate agents. For these reasons, states are prime candidates to bear the next-level responsibility. States become prominent moral agents in circumstances in which valuable ends are not likely to be realized and important moral norms are not likely to be satisfied within a system of social interaction where matters of coordination, distribution of benefits and burdens, and enforcement are beyond the reach of individual agents.

There are, of course, lively disagreements about whether collective action problems are amenable to resolution without state institutional structures or agents vested with coercive power. Moreover, some critics of the prominent role given to states in negotiating and implementing treaties and bearing the costs of addressing the consequences of climate change prefer the backward-looking model of liability under the PPP. Their claim is that all ascriptions of moral responsibility should attach to individuals and corporate entities simply because they are the direct causal agents of environmental problems, rather than governmental entities that are said to have only indirect relationship to the harms (Posner and Sunstein 2008, 20).

However, the uncertainties of effectiveness of non-coercive voluntary solutions, along with the limitations of the liability model, reveal the need for exploration of the clear advantages that states have over individuals as frontline bearers of responsibility for addressing climate change. Under the current global order states possess a unique combination of institutional capacities and political legitimacy for undertaking coercive and definitive resolution of competing claims (Meckled-Garcia 2008). They set the rules for resource ownership, extraction, use, and permissible negative externalities. States shape corporate investment and research and development priorities through tax policy. States shape personal consumption decisions through tax policy and social welfare policies that affect consumer spending. They alter consumption patterns by shaping the distribution of income and wealth and through targeted subsidies for heating oil, solar panels, or employee benefits such as free automobile parking at work. States also create statutorily defined economic incentives and subsidies for production, issue taxpayer-backed bonds that finance industrial construction, and regulate production through a myriad of licensing and accreditation procedures.

In addition, states can counteract the unwanted distributive effects of diffusion of costly new technologies through programs that compensate and retrain displaced employees of fossil fuel-based industries, and help those whose habitat has been adversely affected to adapt to or move to other areas. States thus have at their disposal a wide variety of policy levers that, at least in principle, enable them to bear responsibilities that no other agent can discharge, and they are in position to assign particular responsibilities and burdens to some segments of society with an eye to the demands of fairness and distributive justice.

Further Implications of Morally Unique State Agency

The unique institutional capacities and politically legitimate use of coercive regulatory and coordination power within a state that makes them especially important agents in the division of domestic moral responsibility also explains the moral limits of their role globally. No single nation can avoid climate change on its own, and even a radical reduction of greenhouse gas production by a handful of the planet's largest producers will not be enough to forestall dangerous climate change and prevent serious harm to all of the nations of the world. The upshot is that the outward-facing responsibility of any state is severely circumscribed. The problem of climate change, as well as its solution, is inherently international. Only the creation of transnational institutions, having analogous capacities and politically legitimate authority for coordination, distribution, and enforcement, is likely to be adequate to the task at hand.

That said, states have a unique role in the creation of such institutions. Only states are positioned to create global institutions. The point is not merely the pragmatic observation that within the current global order only an agreement among nation-states can produce binding treaties and new authoritative institutions. The more basic point is a normative one. It is the normative uniqueness of state agency, in both its institutional capacities and political legitimacy of its coercive authority, that enables it to perform according to the terms of any international agreement. No other agent can deliver.

Moreover, the normative uniqueness of states in creating and implementing the terms of a global agreement comes with unique moral constraints on the way such an agreement may be reached. Recall Forst's point about the minimal demands of reciprocity as a constitutive feature of just relations among moral equals. It requires "that none of the parties concerned may claim certain rights and privileges it denies to others." Richard Miller makes a substantially similar point with regard to what citizens of one country, committed to an ideal of reciprocity at the core of universal moral equality, may authorize their own governments to do on their behalf. It requires "backing their own [government's] proposals with morally relevant reasons and giving weight, in proportion to seriousness, to relevantly similar reasons offered by others [governments]" (Miller 2010, 72).

Concretely, because of the morally unique position of governments to negotiate an agreement and their morally unique capacity and authority to implement its terms, reciprocity sets limits on the sacrifices they may demand of the representatives of other nations. They may not press for terms that require sacrifice of the most basic well-being and agency interests of non-nationals unless they are prepared to accept comparable sacrifices among their own citizens. A commitment to a principle of interstate reciprocity does not mean that nation-states cannot drive a hard bargain, or that the pursuit of national self-interest is illegitimate. Nor does it require anything as demanding as tit-for-tat reciprocity. At the very least, however, it means that nations may not prioritize domestic prosperity or a fossil fuel-based, high standard of living for their own citizens at the expense of the most basic, non-negotiable human interests of those hurt first and worst.

Future Generations and Institutional Limitations

Some argue that even the creation of international institutions through a process of interstate agreement remains inadequate to address intergenerational problems of climate change. Another step up the ladder is required. Current generations of interstate bargainers are not reliable stewards of the interests of future generations because of the absence of intergenerational reciprocity. Future generations can do nothing to affect the well-being of current generations, but what current generations do will have enormous impact on future generations. Future generations have nothing to offer the current generation in exchange for their (un-assurable) cooperation. Future generations cannot bargain, and they are powerless to command sacrifice from past generations for the sake of their own well-being. As Stephen Gardiner observes, the intergenerational aspect of the problem makes climate change more resistant to solution. "The usual appeals to broad self-interests rely on there being repeated interactions between the parties where mutually beneficial behavior is possible. But between present and future generations there is neither repeated interaction … nor mutual benefit" (Gardiner 2011, 37).

The intergenerational aspect of the climate change problem is significant. For one thing, as much as 80–90% of the harmful effects of climate change are expected to occur after 2200 (IPCC 2014). Not only are we not "all in it together" internationally: the worst effects will land on some parts of the world even as some parts of the world experience some medium-term benefit from postponing having to reckon with lifestyle changes. It's also the case that generations do not have symmetrical risk-benefit profiles. Current generations have enormous incentive to continue emitting greenhouse gases for their own benefit, and indeed, strong incentives to emit even more in order to raise the standard of living for themselves and several (interacting) generations to come. A significant portion of the world experiences energy poverty, or insufficient energy resources to secure adequate levels of food, water, transportation and medical care. Indeed, the moral focus on the current needs of the poor is built into the original provisions of the United Nations Framework Convention on Climate Change, recognizing the right of development and the differential sacrifices that industrially developed countries should make (UNFCCC 1992).

The implication of enhanced priority for the benefit of the global poor is that those who are charged today with the task of bargaining over the terms of a climate change treaty on behalf of the citizens of their own nations are unlikely to be suitable fiduciaries of the interests of future generations of citizens. There is a built-in conflict of interest, and a high likelihood of bias toward the present, even on the plausible assumption that most people believe that they have responsibilities to those generations not yet born. For these reasons, Gardiner proposes ascent to another level beyond the international institutional level negotiated by self-interested nation-states. Only something like a global constitutional convention can hope to approximate a framework for impartial balancing of interests across generations (Gardiner 2014).

Moreover, arguments for discounting the interests of future generations are offered by economists who assume that future generations will be much better off. William Nordhaus has estimated that those living in 2200 will be 12.3 times richer than those currently living (Nordhous 2006). On the basis of estimates of this sort, it is often claimed that it would be wrong to favor the well-being of future generations over that of current generations who are judged to be comparatively – even dramatically – worse off.

A Substantive Question about What's Owed

Complicating the intergenerational question still further is the inherent difficulty of figuring out what reciprocal fairness across generations would entail. There are lots of unsatisfactory suggestions in play. One thought is that we should aim for overall resource consumption levels that represent a comparable standard of living across the generations. That sort of solution seems intractable. We neither know the size of future generational cohorts nor the specific resource requirements that would result in comparable levels of well-being. We have many reasons to suppose that the planet's population will grow massively. Trying to ensure a comparable standard of living for successive, larger generational cohorts would be too demanding for current generations. Perhaps generations far into the future will have few options but to burn fossil fuels in order to preserve current standards of living. Perhaps they will not. There is considerable uncertainty about what will be required to meet future need, even if we assume that intergenerational fairness demands something like a guarantee of a comparable standard of living.

However, there are some things that we can say with a high degree of confidence. Future humans will have the same basic biological needs. They will require access to food, energy, and water, and for the satisfaction of these biological needs they will need functioning Earth systems. As Henry Shue puts it, if there is one plausible human right it is the right "to inherit from past generations an environment that is neither radically inhospitable nor radically unpredictable" (Shue 2011, 293). Some Earth scientists similarly conclude that what is at stake in climate change is different in kind from all of the rest of what matters in the standard debates about the trade-offs between the environment and other values. They argue that the stable functioning of Earth systems – including the atmosphere, oceans, forests, waterways, biodiversity, and biogeochemical cycles – is a prerequisite above all else (Steffen et al. 2015).

Never before have we had reason to think so modestly about the future, because never before has so much that is so basic been put at such grave risk by the very activities that have contributed so much to human well-being.

Bibliography

Adler, M.D. 2007. "Corrective Justice and Liability for Global Warming." *University of Pennsylvania Law Review* 155 (6): 1859–67.

Baier, Annette. 1985. "Poisoning the Wells." Reprinted in *Postures of the Mind: Essays on Mind and Morals*, 263–291. Minneapolis, MN: University of Minnesota Press.

Broome, John. 1990. "Fairness." *Proceedings of the Aristotelian Society* 91: 87–101.

Broome, John. 2012. *Climate Matters: Ethics in a Warming World*. New York: W.W. Norton and Company.

Caney, Simon. 2010a. "Cosmopolitan Justice, Responsibility, and Global Climate Change." *Climate Ethics: Essential Readings*, edited by S. M. Gardiner, S. Caney, D. Jamieson and H. Shue, 122–145. Oxford and New York: Oxford University Press.

Caney, Simon. 2010b. "Climate Change and the Duties of the Advantaged." *Critical Review of International Social and Political Philosophy* 13 (1): 203–228.

Driver, Julia. 2001. *Uneasy Virtue*. New York: Cambridge University Press.

The Economist. 2009. "India's water crisis: When the rains fail." https://www.economist.com/node/14401149

Forst, Ranier. 2001. "Towards a Critical Theory of Transnational Justice." in *Global Justice*, edited by Thomas Pogge, 169–187. Oxford: Blackwell Books.

Gardiner, Stephen. 2011. *A Perfect Moral Storm: The Ethical Tragedy of Climate Change*. New York: Oxford University Press.

Gardiner, Stephen. 2014. "A Call for a Global Constitutional Convention Focused on Future Generations." *Ethics and International Affairs* 28 (3): 299–315.

Geisler, Charles and Ben Currens. 2017. "Impediments to inland resettlement under conditions of accelerated sea level rise." *Land Use Policy* 66: 322–330. https://doi.org/10.1016/j.landusepol.2017.03.029

Hayward, Tim. 2013. "On Prepositional Duties." *Ethics* 123 (2): 264–291.

IPCC. 2014. "Summary for Policymakers." *Climate Change 2014: Impacts, Adaptation, and Vulnerability. Part A: Global and Sectoral Aspects. Contribution of Working Group II to the Fifth Assessment Report of the Intergovernmental Panel on Climate Change.* C.B. Field, V.R. Barros, D.J. Dokken, K.J. Mach, M.D. Mastrandrea, T.E. Bilir, M. Chatterjee, K.L. Ebi, Y.O. Estrada, R.C. Genova, B. Girma, E.S. Kissel, A.N. Levy, S. MacCracken, P.R. Mastrandrea, and L.L. White (eds.). Cambridge: Cambridge University Press, 2014, www.ipcc.ch/pdf/assessment-report/ar5/wg2/ar5_wgII_spm_en.pdf

Isaac, Morna, and Detlef P. van Vuuren. 2009. "Modeling global residential sector energy demand for heating and air conditioning in the context of climate change." *Energy Policy* 37: 507–521.

Jamieson, Dale. 2007. "When Utilitarians Should be Virtue Theorists." *Utilitas* 19 (2): 160–183.

Jamieson, Dale. 2010. "Ethics, Public Policy, and Global Warming." *Climate Ethics: Essential Readings,* edited by S. M. Gardiner, S. Caney, D. Jamieson and H. Shue, 77–86. Oxford: Oxford University Press.

Jamieson, Dale. 2014. *Reason in a Dark Time.* New York: Oxford University Press.

Kant, Immanuel. 1996. *The Metaphysics of Morals*, edited and translated by M. Gregor. New York: Cambridge University Press.

Kumar, Rahul. 2003. "Who Can Be Wronged?" *Philosophy & Public Affairs* 31 (2): 99–118.

Kutz, Christopher. 2000. *Complicity: Ethics and Law for a Collective Age.* Cambridge: Cambridge University Press.

Laborde, Cecile. 2010. "Republican Global Distributive Justice: A Sketch." *European Journal of Political Theory* 9 (1): 48–69.

Meckled-Garcia, Saladin. 2008. "On the Very idea of Cosmopolitan Justice: Constructivism and International Agency." *Journal of Political Philosophy* 16 (3): 245–271.

Mill, J.S. 1977. *On Liberty*. In *The Collected Works of John Stuart Mill* Vol. XVIII, edited by John M. Robson, 213–310. Toronto: University of Toronto Press.

Miller, David. 2009. "Global Justice and Climate Change: How Should Responsibilities Be Distributed?" *The Tanner Lectures on Human Values* 28, edited by G. B. Peterson, 119–156. Salt Lake City, UT: The University of Utah Press.

Miller, Richard. 2010. *Globalizing Justice: The Ethics of Poverty and Power.* Oxford: Oxford University Press.

Nordhaus, William. 2006. "The Stern Review on the Economics of Climate Change." Available at www.econ.yale.edu/~nordhaus/homepage/homepage/stern_050307.pdf.

Organization for Economic Cooperation and Development (OECD). 2008. *OECD Environmental Outlook to 2030*, Paris: OECD Publishing.

Parfit, Derek. 1984. *Reasons and Persons*. Oxford: Oxford University Press.

Perry, Stephen. 1992. "The Moral Foundations of Tort Law." *Iowa Law Review* 77: 449–514.

Pogge, Thomas. 2007. "Severe Poverty as a Human Rights Violation." in *Freedom from Poverty as a Human Right*, edited by Thomas Pogge, 11–53. Oxford: Oxford University Press.

Posner, Eric and Cass Sunstein. 2008. "Global Warming and Social Justice: What Do We Owe the World for Climate Change?" *Regulation* 31 (1): 14–20.

Powers, Madison. 2015. "Moral Responsibility for Climate Change." in *Routledge Companion to Bioethics*, edited by John Arras, Elizabeth Fenton, and Rebecca Kukla, 133–146. New York: Routledge.

Powers, Madison. 2018. "Sustainability and Resilience." *Encyclopedia of the Anthropocene, vol. 4,* edited by Dominick DellaSala and Michael Goldstein, 29–37. Oxford: Elsevier. http://dx.doi.org/10.1016/B978-0-12-409548-9.10491–9.

Pufendorf, Samuel. 1994. "On the Law of Nature and of Nations." in *The Political Writings of Samuel Pufendorf*, edited by Craig L. Carr, and translated by Michael J. Seidler. New York: Oxford University Press.

Rawls, John. 1999. *A Theory of Justice. Revised Edition.* Cambridge: Harvard University Press.

Ripstein, Arthur. 2006. "Beyond the Harm Principle." *Philosophy and Public Affairs* 34 (3): 215–245.

Ronzoni, Miriam. 2009. "The Global Order: A Case of Background Injustice? A Practice-Dependent Account." *Philosophy & Public Affairs* 37 (3): 229–256.

Sandler, Ronald. 2010. "Ethical Theory and the Problem of Inconsequentialism: Why Environmental Ethicists Should be Virtue-Oriented Ethicists," *Journal of Agricultural and Environmental Ethics* 23 (1): 167–183.

Scanlon, Thomas. 1998. *What We Owe to Each Other.* Cambridge: Harvard University Press.

Scheffler, Samuel. 2010. "Cosmopolitanism, justice, and institutions." Reprinted in *Equality and Tradition: Questions of Value in Moral and Political Theory*, 160–172. New York: Oxford University Press.

Schurer, Andrew P., Michael E. Mann, Ed Hawkins, Simon F. B. Tett, and Gabriele C. Hegerl. 2017. "Importance of the pre-industrial baseline for likelihood of exceeding Paris goals." *Letters: Nature Climate Change* 7: 563–567. www.nature.com/natureclimatechange

Shue, Henry. 2010. "Subsistence Emissions and Luxury Emissions." in *Climate Ethics: Essential Readings*, edited by S. M. Gardiner, S. Caney, D. Jamieson, and H. Shue, 200–214. Oxford and New York: Oxford University Press.

Shue, Henry. 2011. "Human Rights, Climate Change, and the Trillionth Ton." in *The Ethics of Global Climate Change,* edited by Denis Arnold, 292–314. Cambridge: Cambridge University Press.

Sinnott-Armstrong, Walter. 2010. "It's Not My Fault: Global Warming and Individual Moral Obligations." in *Climate Ethics: Essential Readings*, edited by S. M. Gardiner, S. Caney, D. Jamieson, and H. Shue, 332–346. Oxford and New York: Oxford University Press.

Steffen, Will, Katherine Richardson, Johan Rockström, Sarah E. Cornell, Ingo Fetzer, Elena M. Bennett, Reinette Biggs, Stephen R. Carpenter, Wim de Vries, Cynthia A. de Wit, Carl Folke, Dieter Gerten, Jens Heinke, Georgina M. Mace, Linn M. Persson, Veerabhadran Ramanathan, Belinda Reyers, and Sverker Sörlin. 2015. "Planetary boundaries: Guiding human development on a changing planet." *Science* 347 (6223). http://dx.doi.org/10.1126/science.1259855

United Nations Framework Convention on Climate Change (UNFCCC). 1992. http://unfccc.int/files/essential_background/background_publications_htmlpdf/application/pdf/conveng.pdf.

United Nations Framework Convention on Climate Change (UNFCCC). 2015. *Adoption of the Paris Agreement.* http://unfccc.int/resource/docs/2015/cop21/eng/l09.pdf

Wenar, Leif. 2007. "Responsibility and Severe Poverty." in *Freedom from Poverty as a Human Right: Who Owes What to the Very Poor?,* edited by Thomas Pogge, 255–274. Oxford: Oxford University Press.

3

PROCREATION, CARBON TAX, AND POVERTY

An Act-Consequentialist Climate-Change Agenda

Ben Eggleston[1]

Introduction

In 2009, the U.S. government implemented the "Cash for Clunkers" program, providing rebates for replacing fuel-inefficient vehicles with relatively fuel-efficient ones. The program was popular among consumers, with nearly 700,000 of them exhausting the allocated funds in a matter of weeks. There were other benefits, too, including reduced greenhouse-gas (GHG) emissions, reduced emissions of traditional pollutants, and aid to automobile manufacturers who had been hit hard by the recession that had begun two years earlier. On the other hand, the program had costs, such as US$3 billion in public funds, the destruction of drivable cars worth hundreds of millions of dollars (car dealers were required to ruin the engine of every qualifying "clunker" traded in, regardless of the car's actual functioning), the resultant shortage of inexpensive used cars for sale, and the increased GHG emissions attributable to the increased manufacture and transport of new cars, due to program-induced demand (Lenski et al. 2013, 173).

To evaluate this program and other attempts to slow the pace of climate change, one option is to consider what seem to be the most salient facts and form an intuitive judgment. Another option is to adopt the perspective of a particular moral theory and apply it to problems and decisions that relate to climate change. This latter approach could be useful both in generating a more probing and systematic assessment of the programs and policies under consideration, and in constituting an extensive testing ground for the theory itself. As an instance of this approach, this chapter is devoted to examining the ethics of climate change through the lens of act consequentialism.

Act Consequentialism

A Principle of Rightness

The eighteenth-century religious thinker John Wesley is said to have urged his followers to live their lives this way:

> Do all the good you can, by all the means you can, in all the ways you can, in all the places you can, at all the times you can, to all the people you can, as long as ever you can.
>
> *Shafer-Landau 2018, 120*

This is a paradigmatically act-consequentialist piece of advice, since making the world as good as it can be is the basic idea of act consequentialism.

As its name suggests, the view is often articulated in terms of consequences – specifically, the consequences of acts. To be precise, any moral view that includes the following principle is a form of act consequentialism:

> An act is right if and only if its consequences are at least as good as the consequences of any act the agent could have performed.

Act consequentialism is sometimes understood in terms of the *states of affairs* that result from actions: An act is right if and only if it results in the best available state of affairs. Of course, the notion of a state of affairs needs to be understood as temporally extended into the indefinite future, not as a snapshot of the world immediately after the action is performed. Short-term pain followed by long-term comfort is a better state of affairs (in this temporally extended sense) than short-term comfort followed by long-term pain.

Assumptions about the Good

Act consequentialism per se is simply a thesis about the absolute and direct dependence of the right on the good: It says that acting rightly is a matter of maximizing the good. In this minimal form, this view is neutral about what the good is, and is compatible with any view about the good that one may wish to hold. A moment ago, I assumed that comfort is better than pain, but one could be an act consequentialist and hold that pain is better than comfort, or that they are equally good, or that they cannot be compared.

Moreover, in order to avoid being practically vacuous, act consequentialism *must* be conjoined with some view about the good, in any serious attempt to analyze a real-world ethical issue from an act-consequentialist point of view. Accordingly, in this section I will put forward some assumptions about the good that will be operative in this chapter. These assumptions will be among the least controversial options that a theorist could choose in order to flesh out a sufficiently determinate act-consequentialist moral theory, though I do not imagine any of them to be beyond debate.

The well-being of sentient creatures

I take it that the most important constituent of the goodness of states of affairs is the well-being of human beings and other sentient creatures. This, of course, raises the question of what constitutes a sentient creature's well-being, with the added complication that the answers might differ among types of sentient creatures. For the purposes of this chapter, we can take an ecumenical approach, stipulating that the well-being of a sentient creature (whether human or otherwise) is constituted by some combination of the pleasantness of its conscious experience, the satisfaction of its desires, and its possession of certain objectively good things such as meaningful relationships with others (though obviously the applicability of these criteria will vary among different types of creatures). In some contexts, this approach would be unacceptably indefinite. But here, its pliability averts unnecessary controversy, and the unresolved issues are likely to be moot in in the context of this chapter. For example, climate-change debates rarely pivot on whether people's lives should be pleasant even if devoid of objective goods, or vice versa.

Total well-being and its distribution

Some people who value well-being, as just proposed, believe that the greater the total quantity of well-being, the better. Although I endorse that principle myself, it is one of the more extreme views on this topic. More popular, I think, is the idea that while total well-being matters, some facts about its distribution matter, too. For example, many people think that if well-being is distributed highly unequally in some state of affairs, then some (not too large) decrease in total well-being would, morally, be a price worth paying for some (not too small) increase in the quantity of well-being enjoyed by the worst-off. These unresolved issues deserve continued attention. As it happens, however, the arguments of this chapter will be sufficiently general as to be largely neutral with respect to them.

Goods beyond well-being

Although the view about the good sketched so far is inclusive in certain ways (e.g., by allowing that the well-being of many nonhuman animals matters, morally), it does not specifically affirm the value of anything other than well-being. One might, however, ascribe value to rational agents' having opportunities to exercise their autonomy, or to the preservation of naturally occurring biodiversity, or to other things beyond and separable from well-being. Here I take no stand on such further possible constituents of the good, other than to assume that they are, at best, much less important than well-being.

No temporal discounting

Many of the costs and benefits of climate change will occur far in the future. In some disciplines, such as traditional cost-benefit analysis, it is customary to give future costs and benefits less weight than present ones. This is done by "discounting" costs and benefits in proportion to their remoteness, typically at a rate of about 5 percent per year (Gardiner 2011, 267). For example, if a particular harm (say, a major hurricane passing through Houston) will occur either now or twenty years from now, the latter scenario is regarded as being $(0.95)^{20}$, or 36 percent, as bad as the former.

Although such discounting is customary in certain fields, it is "the most controversial issue in climate economics" (Gardiner 2011, 267). Here I reject temporal discounting because of its confused conceptual foundations, often sloppy implementation, and ethically suspect implications. To take the last point first, any nontrivial discount rate implies that we can virtually disregard the interests of all but the nearest future generations. For example, if we were making a policy decision that could affect whether Mumbai would become uninhabitable a century from now (say, due to higher sea levels), a 5-percent discount rate would imply that, in our deliberations, we should regard that event as being less than 1 percent as bad as if it were to occur today, and that we should prefer the century-delayed loss of up to 100 such cities over the loss of one such city today.

Proponents of temporal discounting give many conceptual rationales for it: temporally remote events are less certain, future people will be richer, resources invested now will be worth more later, the value of money tends to decrease over time due to inflation, and others. But as Tyler Cowen and Derek Parfit point out in their sharp critique of such rationales, these considerations are all distinct from *temporal* discounting (Cowen and Parfit 1992, 145–150). For example, if a remote event is less certain, we discount it for its uncertainty, not its remoteness. Similarly, if some resource can go to a rich future person or a poor present person, we discount the former scenario because of the diminishing marginal utility of money (or because we believe benefits to the rich are worth less, morally, than same-size benefits to the poor), not because the rich person is in the future. Regardless of the merits of discounting for greater uncertainty, greater wealth, and other factors, these are not rationales for temporal discounting. At best, these factors correlate with remoteness in time. But as Cowen and Parfit point out,

> Remoteness in time correlates with a whole range of morally important facts. So does remoteness in space. Those to whom we have the greatest obligations, our own family, often live with us in the same building. We often live close to those to whom we have other special obligations, such

> as clients, pupils, or patients. ... But no one suggests that, because there are such correlations, we should adopt a spatial discount rate. ... The temporal discount rate is, we believe, as little justified.
>
> *Cowen and Parfit 1992, 159*

Conspicuously lacking from the temporal-discounting literature is the claim that temporal remoteness is itself a morally important fact – perhaps indicating the indefensibility of that thesis as well as the confusion clouding the proffered rationales.

A dogged cost-benefit analyst might acknowledge these conceptual distinctions but hold that, as a matter of implementation, remoteness in time is a good proxy for morally important facts such as greater uncertainty and greater wealth. Obviously such a claim can be assessed only in light of the details of its proposed application, but it warrants skepticism because of the many ways the alleged correlations can break down. Events are unpredictable not just because of remoteness in time but also because of other factors such as novel circumstances, poorly understood processes, and the dynamics of chaotic and complex systems. The latter factors can reverse the usual correlation between nearness in time and predictability: for example, it is probably easier to plausibly predict the hydroelectric output of the Colorado River two generations from now than U.S. economic output two decades from now. Similarly, using temporal discounting as a proxy for rising wealth is undermined by the fact that today's extreme economic inequality shows no sign of diminishing; consequently, many of the world's future people will look back with envy on the living conditions of many of the world's present people.

The literature on discounting is vast, and here I must pass over some potentially important views, such as Cowen's argument that the rejection of temporal discounting is not merely defensible but, rather, logically entailed by the axioms implicit in any moral theory "that attempts to evaluate and compare outcomes" (Cowen 1992, 162). Here I have just aimed to say enough to explain and motivate my decision to exclude temporal discounting from the conception of the good assumed here.

Objections

Although the idea of making the world as good as it can be has obvious intuitive appeal, there are two broad objections to act consequentialism that deserve to be mentioned in this overview of the theory.

First, the theory is difficult to implement. The most obvious way of attempting to apply the theory to an ordinary decision-making situation is to identify the acts that can be performed in that situation, ascertain the state of affairs that would result from each act, identify the best of those states of affairs, and choose the act that causes that state of affairs. But for many possible acts, the predictions that can reasonably be made are rough and conjectural, at best. For example, it is typically far from certain what will be the consequences of a high-school senior's selection of a particular college to attend, a parole board's decision to grant early release to a particular prison inmate, or a political party's decision to select a particular candidate as its nominee for president.

Second, the theory has implications that clash with what many people take to be basic truths of morality. For example, it implies that promises ought to be broken, that friends and family members ought to be betrayed, and that individual rights ought to be violated, whenever doing so would produce sufficiently large benefits. Similarly, act consequentialism seems to require extreme self-sacrifice on the part of many people who, by giving up much of their personal property (including their internal organs), could benefit other people more than they would be harming themselves. Thus, act consequentialism seems to conflict with many of the prohibitions and many of the permissions of common-sense morality.

In principle, a defender of act consequentialism as a principle of rightness could conjoin it with a theory of the good specially tailored to make the resulting theory less vulnerable to the foregoing objections. For example, to address the prediction problem, one might hold that sufficiently remote effects are neither good nor bad – at least, for the purposes of evaluating acts. And to address the problem of immoral implications, one might hold that breaking promises, betraying friends and family, and violating individual rights are grave bads, while the protection and exercise of considerable autonomy with respect to personal property is a major good.

But such ad hoc fixes are unpromising, for several reasons. First, they create new problems: the devaluing of remote effects is an extreme form of temporal discounting, and it may address the prediction problem at the cost of making the revised view even more substantively implausible than ordinary forms of temporal discounting are. And any elevating of individual rights is likely to make the revised view vulnerable to new objections about its relative neglect of non-rights-related goods. (Consider, for example, how a strongly formulated right of self-ownership can preclude the taxation necessary for basic government services.) Second, such fixes frequently work at cross-purposes: consider, for example, the revised view's apparent ambivalence about temporally remote violations of individual rights – Are they unimportant because they are remote, or important because they are violations of individual rights? Also, the more promiscuously a theory designates various goods and bads as *especially important*, the less of a difference that designation makes in the theory's comparisons of alternative possible states of affairs. (When everything is special, nothing is.) In the end, it seems unlikely that these objections to act consequentialism can be avoided to any meaningful extent by way of a carefully formulated conception of the good.

There are, however, other replies to these objections that are often regarded as having some merit. In regard to the prediction problem, two points are commonly mentioned. First, act consequentialism is a theory about what makes acts right and wrong, not a decision procedure for selecting an act in any given situation. The difficulty of predicting possible acts' consequences – which is simply a fact about the natural world and our ways of knowing about it – does not preclude or impinge on the *value* judgment that those consequences are what make those possible acts right or wrong. Second, moral perspectives other than act consequentialism (whether they are other theoretical perspectives or simply common-sense morality) almost invariably regard consequences as being important to the rightness and wrongness of acts. (They just tend to differ from act consequentialism in holding that other aspects of acts matter, too.) But the prediction problem is not seen as vitiating the consequentialist strands of those other perspectives – and rightly so, since people make prediction-based decisions constantly and in all facets of life without feeling that they are tying their fate to the output of an epistemically impossible mental process. On reflection, then, the prediction problem is hardly the special burden for act consequentialism that it might initially seem to be.

The second objection – the problem of immoral implications – can be addressed here only in the most cursory fashion, because of the variety of the topics it comprises. Ultimately, though, most replies to instances of this objection are deployments of a few recurring argumentative strategies. One is to emphasize that act consequentialism prescribes breaking promises, violating individual rights, and other prima facie immoral behavior *only* when such conduct would produce *sufficiently large benefits* – a critical proviso. A second strategy is to point out that breaking promises, violating individual rights, and so on tend to have very bad consequences, not only through their direct effects in individual instances but also indirectly, by eroding background conditions of trust and security. Like bodily health, these background conditions are easy to take for granted and may not be foremost in one's mind until they start to break down. But in any rational consequentialist assessment of various kinds of acts, it would be shortsighted not to acknowledge the grave (albeit often remote) long-term effects of breaking promises, violating individual rights, and so on. A third and final strategy for responding to objections – quite different from the two strategies just discussed – is to cast doubt on the reliability of the nonconsequentialist intuitions that the objections rely on for their dialectical force. A standard way of doing this is to argue that our moral intuitions were largely shaped by forces of natural selection that operated in an era in which most people lived in small tribes. Reproductive fitness was not enhanced by character traits such as being concerned about the impact of one's actions on far-away or future people, or simply being inclined to see oneself as no more important, morally, than anyone else. Of course, there is much more to be said about this and the previous strategies for responding to objections to act consequentialism, but this will conclude our overview of the theory.

The Default Moral Theory of Climate Change

Before applying act consequentialism to the ethical issues raised by climate change, it is worth noting the extent to which something like act consequentialism is already presupposed by scientists, environmental advocacy groups, public-policy professionals, and politicians in their discussions of climate change. I have to say "*something like* act consequentialism" because very few people are thoroughgoing act consequentialists. I do claim, however, that if we were to try to interpret the moral concerns animating the global conversation about climate change as expressions of some moral theory or other, we would find act consequentialism to be a more natural fit than, say, rule consequentialism, contractualism, or Kantianism. Consider, for example, the following questions:

1. In light of probable ongoing and future climate change, what acts that we can perform will have the best consequences?
2. In light of probable ongoing and future climate change, what are the rules whose general internalization would have the best consequences?
3. In light of probable ongoing and future climate change, what rules would be agreed upon by persons deliberating in an appropriately characterized initial situation?
4. In light of probable ongoing and future climate change, what maxims for action are universalizable?

If one were to survey contemporary discussions of climate change – ranging from experts' technical treatises to informal conversations around the dinner table – one would find trains of thought far more aligned with the first question than with any of the others. Thus, it seems fair to regard act consequentialism as the *default moral theory of climate change*, and to think that inquiries of the kind conducted in this chapter – exploring the implications of act consequentialism for climate-change ethics – would be especially pertinent to our current predicament.

To be sure, the moral perspectives underlying most discussions of climate change do deviate from act consequentialism in one major way, by focusing on preventing or lessening the *harms* of climate change rather than attending to the broader mandate of making the world as good as it can be. However, this is only to be expected in light of the fact that climate change seems to present us with many problems – many big problems – and very few opportunities. Thus, the focus on harms is most likely a matter of emphasis, arising out of the grim reality of climate change, rather than a repudiation of the idea of making the world a better place.

It might be pointed out that most people actually reject act consequentialism – for reasons having to do with the objections mentioned above – and it might be asked how a theory that most people reject could be called the default moral theory of anything (whether climate change or otherwise). It turns out, however, that those objections do not have much relevance to climate-change ethics, since it is exceedingly rare for people to have the opportunity to substantially improve the climate by, say, breaking a promise or violating someone's rights. Moreover, most non-philosophers (and many philosophers) do not apply a single moral principle to every issue they think about. Instead, they are content to use different frameworks in different contexts. And in the context of large-scale, long-term problems such as climate change, many people seem to operate with frameworks that align fairly closely with act consequentialism. Robert Goodin persuasively argues that utilitarianism "can be a good guide to public affairs without its necessarily being the best practical guide to personal conduct" (Goodin 1995, 4). Given the similarities between utilitarianism and the well-being-focused form of act consequentialism presented here, Goodin's distinction might explain why many people who decline to endorse act consequentialism as a principle, or who disagree with its implications in some situations, still seem to engage in essentially act-consequentialist forms of reasoning when considering the ethics of climate change.

Personal and Policy Recommendations

The remainder of this chapter will be concerned with discussing possible responses to the threat of climate change. I will discuss just three topics, though each is a realm in which any decisions made (even by omission) are enormously consequential. I start with a topic in the domain of personal morality – whether people should have fewer children, in order to reduce the total number of future people emitting greenhouse gases. I then turn to two topics in the domain of public policy. The first, taxing GHG emissions, is standard fare in discussions of climate change, but its merits cannot be repeated often enough (until it is enacted). The other one, concerning poverty, is currently somewhat less prominent, but also deserves to be at the center of climate-change ethics.

I consider just one topic of personal morality before turning to policy-level recommendations because policy-level changes are essential to effectively addressing climate change. As Peter Singer writes, actions such as not eating meat and driving less "are good things to do, but we should not fool ourselves into believing that the problem of climate change can be solved by individual actions of this kind. There need to be changes on a larger scale" (Singer 2016, 26). Nevertheless, the procreation topic also deserves careful inquiry, because of both the enormous ramifications of any decisions made in that realm and the difficult theoretical challenges that must be met in order to formulate defensible principles for making such decisions.

Procreation

Most recommendations for reducing GHG emissions pertain to sacrifices and efforts that people can make throughout their lives, in greater or lesser degrees: using less heat and air conditioning, driving less, recycling more. Underlying the rationales for these lifestyle adjustments is the fact that the life of the typical human being – who chooses among the ordinarily available options for housing, transportation, occupation, recreation, and so on – tends to make the climate worse. Now, the non-climatic moral reasons bearing on the advisability of ending the life of any already-existing person are typically so weighty that it seems unfathomable that climate-change considerations could significantly influence the rightness or wrongness of the intentional death or continued life of such a person: no one should commit murder or suicide just so there's one fewer GHG emitter walking around. But even if we regard the termination of lives as beyond the bounds of climate-change ethics, we should consider the implications of climate-change considerations for choices about the creation of lives – specifically, prospective parents' choices of whether to have children.

Because lifestyle adjustments (such as turning down the thermostat) only marginally reduce a human being's lifetime GHG emissions, the choice of not having a child – or having one fewer child than one otherwise would – has effects at a whole other order of magnitude. As Dale Jamieson writes,

> If an American wants to minimize his environmental impact, the most effective thing he can do is to refrain from having children. He can drive around in an SUV, hang out at McDonald's, take long hot showers and still have much less environmental impact than if he fathers one, good, green, nature-loving American child.
>
> *Jamieson 2008, 189*

A recent study quantifies the impacts of about a dozen different GHG-reducing actions more precisely. In this study, perhaps the most striking statement is that "a US family who chooses to have one fewer child would provide the same level of emissions reductions as 684 teenagers who choose to adopt comprehensive recycling for the rest of their lives" (Wynes and Nicholas 2017a, 3). But to be honest, although the number 684 is impressive, I'm underwhelmed by the choice of recycling as a reference point, and focusing on a U.S. couple inflates the figure since U.S. per-capita GHG emissions are among the highest in the world. For me, the most meaningful comparison derivable from the study's computations is that if a couple in the *developed world* (not just the U.S.) were to decide, at some point in time, to have one fewer child, then the GHG-reducing effect of that decision would be about *nine times* as large as the *combined* GHG-reducing effect of their doing the following *for the rest of their lives*: adopting comprehensive recycling, not eating meat, buying green energy, getting rid of their two cars, and taking two fewer transatlantic flights per year (Wynes and Nicholas 2017b, 6).

Thus, having fewer (or no) children is an extraordinarily effective way for people to reduce GHG emissions. But we cannot ascertain what act consequentialism implies about the morality of having children without considering the other consequences of having children. Obviously those other consequences are numerous and varied, but here I want to focus on just two. First, for most people, the number of children that they have is among the most life-defining decisions that they will ever make. Whether one wants several children, or just one, or none at all, one is likely to feel very strongly about that preference, and its fulfillment or frustration is likely to be a major determinant of one's well-being. Since the well-being of the agent counts in an act-consequentialist assessment of the morality of an action, a person's desire to have a child, or a certain number of children, must be recognized as a sizable justifying consideration.

The other major consequence that needs to be considered is, simply, the existence of the child as a sentient creature with, let us assume, a life worth living. And here we confront what may be the most fundamental question of population ethics: Do more people – assuming they have lives worth living – make the world a better place? Within this chapter's act-consequentialist framework, there is an obvious route to an affirmative answer to that question: If there are more people with lives worth living, then there is more well-being, and the world is a better place. On this view, when new people come into existence, there may be disvalue because of GHG-related considerations, but there may also be value because of the individuals' personal well-being – and the latter might exceed the former. And when that is the case, refraining from having children might make the atmosphere better, but make the world as a whole worse. It must be admitted, however, that this fundamental question of population ethics is controversial, and answers contrary to mine also have powerful reasons supporting them.

Where, then, do we stand? Having children is bad for the environment, but is often important or even essential to their parents' well-being. Also, bringing into the world more people with lives worth living is good in itself – or maybe it's not. Finally, all of these issues are implicated not just in personal morality, but also in many areas of public policy. Perhaps the clearest conclusion we can draw here, on the topic of having children as an aspect of climate-change ethics, is that resolving its conundrums is likely to be as difficult as it is important.

Taxing GHG Emissions

Probably the single most effective step that could be taken to responsibly reduce GHG emissions would be the implementation of a tax on such emissions – commonly called a carbon tax, though it would apply to all major greenhouse gases in proportion to the harmfulness of their effects. Ideally, this would be done on a global scale, with the participation of all of the countries of the world. But it could also be done regionally, such as by the European Union, or just by a single country such as the United States.

Let me sketch the basic idea of a carbon tax. In theory, it would be a tax paid whenever anyone engaged in an activity that caused GHG emissions. For example, if a power plant burns coal to generate electricity, then it would pay a tax proportional to its GHG emissions. Similarly if an airline burns jet fuel to fly from point A to point B, or if a livestock firm maintains a warehouse full of cows emitting methane. The main purpose of such a tax would not be the usual purpose of taxation – generating revenue – but making GHG-emitting behavior more costly and, therefore, reducing its occurrence. (It would be a Pigovian tax, in the parlance of economics.) Such behavior changes would be expected because if it becomes more expensive for firms to use high-emissions methods of producing their output (whether electricity, transportation, or beef), then there will be substitution effects, both in production and in consumption. At the production stage, the tax will incentivize firms to switch to lower-emissions ways of producing the same product – whether by retrofitting a power plant, buying more fuel-efficient planes, or (somewhat more speculatively) making beef in a petri dish rather than in an animal. At the consumption stage, to the extent that firms end up with higher production costs and have to raise their prices, the higher prices will cause some consumers to change their purchasing decisions. Without assuming anything about the behavior of any given individual, we would expect to see, over the whole population affected, small decreases in the consumption of electricity, air travel, and beef.

I mentioned that, in theory, a carbon tax would be paid whenever anyone engaged in an activity that caused GHG emissions. Taken to its logical extreme, however, this would have absurd implications. For example, it would imply that just by being alive for another day (and emitting carbon dioxide every time I breathe), I should pay a tax, with perhaps a surcharge if my respiration needs are increased by being out of shape or exercising a lot. (So, they get me either way.) In practice, however, it would be possible to exempt such activities altogether, and focus on industrial production as in the examples above. Such simplifications would not have major distortionary effects, since not many people would, for example, exert themselves inefficiently generating their own electricity on stationary bikes because they want to avoid paying for industrially produced electricity.

Another system for discouraging GHG-emitting behavior is known as "cap and trade." Under this system, governments "cap" aggregate GHG emissions for a certain interval of time at a certain level (based on policy judgments about the social costs and benefits of various levels of emissions) and allocate emissions permits to major emitters such as power plants. The permits can be bought and sold – this is the "trade" part of the system – enabling firms that can reduce their emissions to make money by selling their unused permits to firms who find that buying permits is cheaper than reining in their emissions. Because firms profit from emissions reductions and pay more for overages, they have the same incentives under a cap-and-trade system as they do under a carbon tax. In fact, as the economist William Nordhaus explains, the two systems "are fundamentally the same. That is, in an idealized situation, they have the same effects on emissions reductions, on carbon prices, on consumers, and on economic efficiency" (Nordhaus 2013, 237). In practice, however, the systems sometimes operate differently; I choose to focus on a carbon tax for two reasons given by Nordhaus: taxation is a more established and familiar mechanism, compared with cap and trade; and in practice, the efficacy of cap-and-trade systems has been compromised by extreme volatility in the price of permits (Nordhaus 2013, 239). However, cap-and-trade systems also have advantages over tax systems (Nordhaus 2013, 240), and discouraging GHG-emitting behavior is such an important policy goal that either instrument is far better than neither one (Nordhaus 2013, 241).

Implementing a carbon tax would solve three major problems that arise from conventional environmental regulations. Currently, most developed countries have vast patchworks of environmental regulations, with distinct sets of rules for the emissions of power plants, the fuel efficiency of cars, the efficiency of air conditioners, the efficiency of light bulbs, and countless other industrial practices and consumer products. There is also a separate patchwork of grants for purposes ranging from basic research on carbon sequestration to more insulation for old homes. One problem with this patchwork approach is that it relies on a case-by-case determination, by government agencies, of what amount of environmental benefit is worth what amount of increased costs of production and consumption (or what amount of government funds, in the case of the grants). This approach involves heavy reliance on the technical expertise and moral judgment of government agencies (many of which are headed by political appointees rather than career professionals). In contrast, the carbon-tax approach would let efficiencies in every sector of the economy emerge in response to financial incentives, and many regulations could be taken off the books. For example, it would be unnecessary for government regulators to declare inefficient light bulbs illegal, since sales of them would drop in response to consumers' increasing sensitivity to their own electricity consumption. (If a small quantity continued to be sold, to meet either special needs or quirky preferences, that might even be better, from an act-consequentialist point of view, than banning them altogether.) And power plants would be incentivized to invest in carbon-sequestration research – unless their experts thought it was a pipe dream, in which case maybe the government shouldn't be giving grants for it anyway.

A second problem with the patchwork approach is that it is almost impossible to implement with a consistent standard of how much environmental benefit is worth how much financial cost. Federal agencies are required to consider a centrally determined "social cost of carbon," but considerable scope for discretion remains (Malakoff et al. 2016, 1365). Given the vagaries of human psychology (and the particular corruptions that typically hobble regulatory agencies, such as being captured by lobbyists), it is almost certainly the case that some environmental regulations are much more stringent than others, and that actors in some sectors of the economy are being asked to make sacrifices far out of proportion to the sacrifices being required of actors in other sectors of the economy. For example, given that industries tend to have better lobbyists than consumers do, it is likely that many industrial activities are held to lower standards of environmental concern than are implicit in many consumer-products regulations. Replacing the patchwork approach with a carbon tax would cut off these unfair inconsistencies at their source, and replace them with the same standard for everyone.

A third problem with the patchwork approach is that, despite the enormous amount of information that it requires for its input (as I mentioned two paragraphs ago), it produces a frustrating dearth of information among its outputs. Specifically, it provides very little information for people to use when trying to make choices in a GHG-minimizing way. Suppose, for example, that I plan to go to Chicago, and I am deciding whether to drive, fly, or take the train. Suppose also that I care about ascertaining which option would minimize GHG emissions, either because I am otherwise indifferent among the options or because I value the environment enough to incur some extra cost for it. The existing system of regulations does not put me in a position to know anything more than that each option is not so bad for the environment that it's *illegal*. I might be able to do some research about the GHG effects of the different modes of transport and apply that research to the specific details of my trip, but that would quickly get unwieldy. In contrast, a rationally determined carbon tax would more closely align the prices of the different options with their GHG emissions. There would not be perfect alignment, since the prices would reflect the fundamental production costs as well as the carbon tax. But even that complication is desirable from an act-consequentialist point of view, because people should be discouraged from consuming a particular good or service not only in proportion to its GHG emissions, but also in proportion to its use of physical goods, the time and expertise of human beings, and so on.

One final aspect of a carbon tax that I want to discuss is what to do with the money that is collected. I mentioned above that the main goal of a carbon tax is to change behavior, not to raise money. But unless the tax is set so high as to make all GHG-emitting activities utterly cost-prohibitive – which is not a realistic possibility – there will be tax revenue, and decisions will have to be made about what to do with it. Obvious candidates include reducing the debt, reducing taxes, and funding government services (see Kestenbaum 2013 and Kestenbaum et al. 2018, which also touch on several other aspects of carbon taxation discussed here). Here I want to mention another option – namely, just dividing the total revenue by the total number of adults in the country and sending each of them a rebate check for that amount at some convenient interval, such as annually. Every January 31 (or whatever), the government would announce the rebate for the previous year, and every adult would get a check for that amount.

This proposal has some similarities to one of the unique features of living in Alaska: the Permanent Fund Dividend, under which proceeds of some of the state's mineral revenues (especially oil revenues) are invested in a state-owned fund which then generates dividends paid directly to the residents of the state. The annual amount varies, but is typically on the order of $1,000 (Griffin 2012, 79). A carbon-tax proposal in the same vein is unorthodox, but it has several advantages over other options. First, it would blunt the criticism – sure to be mounted against any carbon tax – that despite the rationale of trying to discourage harmful behavior, it is really (just like all other taxes) just another way for bureaucrats to grab more money from hard-working citizens. Second (and relatedly), it would make the tax more politically viable to implement, and politically more difficult to eradicate. Third (again, relatedly), it would pave the way for occasional increases in the carbon tax, which might be desirable either as part of an initial phase-in of the tax or in response to scientific and economic findings indicating that the tax is, at some time in its operation, too low. Fourth – turning now from political considerations to directly moral ones – the rebate program would help low-income people cope with the financial impact of paying more for gasoline, electricity, and other necessities that would become more costly under the tax. Fifth, it would lessen economic inequality, by taxing wealthy people more (because they have higher GHG emissions) but not giving them greater rebates. Sixth, and finally, it would promote a national sense of solidarity, by representing the revenue of the carbon tax as a joint resource from which all should benefit equally.

It might be thought that if the revenue is just returned to the citizens every year, then the financial impact will be neutralized and no one will be motivated to change their behavior away from high-GHG activities. It is true that the *collective* financial impact will be neutralized, but each *individual* will still have virtually the same incentive to choose low-emissions activities over high-emissions ones. If my city told me it was going to institute this kind of rebate program to disburse the proceeds of parking tickets every year, it would still be in my interest to avoid getting parking tickets. And likewise with a carbon tax. (A few sentences ago, I had to say "*virtually* the same incentive" because when people feel wealthier – which would be an effect of the rebate program – they tend to be a little less responsive to price differences. But this "wealth effect" on consumer choice would be negligible.)

In the preceding paragraphs, I have indicated some of the benefits of an egalitarian rebate program. Of course, in terms of the overall framework of this chapter, the pertinent question is whether these benefits are great enough to make such a program have better consequences than other possible uses of the funds. As I mentioned, other possible uses include debt reduction, tax cuts, and government services. It would be far beyond the scope of this chapter to survey the pros and cons of all such options and pretend to convincingly identify the best one. There is, however, one objection to the egalitarian rebate program that I want to address. It is sometimes suggested that the rebate should be more redistributive than I propose, with most or all of the funds going to the poor, and that the act-consequentialist rationale for preferring this over an egalitarian program is as clear as the act-consequentialist rationale for preferring a progressive income tax over a regressive income tax. Now, I would acknowledge that if, say, a trillion dollars were to rain down one time on the people of a country such as the United States, the consequences would be better if each person in the bottom quintile of wealth were to receive $15,000 (with everyone else receiving nothing) than if each person in the country, rich and poor alike, were to receive $3,000. But any carbon tax will face constraints of political feasibility, both when proposed and when attacked after enactment, and I think that any non-egalitarian proposal for a rebate program runs a much greater risk of being successfully tarred as just another big-government income-redistribution program. Of course, feasibility is a matter of degree, and in American society currently, any national carbon-tax proposal would be dead on arrival in the U.S. Congress (unless, perhaps, the funds were designated to go straight to the oil companies). But my proposal of an egalitarian rebate program is meant to show how considerations of producing optimal consequences can appropriately be tempered by at least some concessions to political feasibility.

Poverty Reduction

The last topic I want to discuss – poverty reduction – is rarely regarded as a priority of climate-change ethics. Of course, poverty reduction has been a major topic of the broader field of applied ethics at least since the 1972 publication of Peter Singer's article "Famine, Affluence, and Morality." But it is not typically a focal point of discussion for people concerned about the effects of climate change. Indeed, one can easily get the impression, from contemporary philosophical and popular discussions, that poverty reduction and climate change are two separate areas of applied ethics. For example, Thomas Schelling characterizes the relationship between slowing climate change and aiding the poor in terms of "trade-offs" – one against the other (Schelling 1997, 8). Here, however, I argue that reducing poverty should be seen as one of the central imperatives of climate-change ethics.

The standard narrative in climate-change ethics is that we are spoiling the planet for our children and grandchildren. Stated a little more explicitly, the idea is that ever since the industrial revolution or so, our atmosphere has been on a dangerous warming trajectory. For a while – maybe the entire nineteenth century or so – people could be excused for not knowing about the greenhouse effects of burning coal and other fossil fuels. But scientific knowledge of the greenhouse effect (and public dissemination of that knowledge) advanced tremendously during the twentieth century, and it has been at least several decades since we last had excusable ignorance of the consequences of our actions. Yet we continue to increase our GHG emissions every year, wrongly leaving our descendants an atmosphere much worse than the one we've enjoyed. Hence the title of the first major chapter of Stephen Gardiner's contribution to *Debating Climate Ethics*: "Betraying the Future" (Gardiner and Weisbach 2016, 6).

Intergenerational justice is certainly an important dimension of climate-change ethics. But the story is not as simple as it might seem. One complication – though a very fortunate one overall – is that in many ways the world we will pass along to our descendants will be better than the one we inherited. There are many resources other than the natural environment that can contribute to people's well-being, including infrastructure such as wells, sewers, roads, and ports; well-functioning institutions such as legal systems, financial systems, and school systems; useful professions such as medical science, engineering, information technology, and journalism; cultural artifacts such as television, film, and literature; and morals and norms that restrain unethical behavior. Given these many dimensions along which current people can help or hurt future people, it seems likely that the requirements of intergenerational justice have more to do with the totality of them than just one or two of them such as the temperature of the atmosphere or the level of the sea. And along many of these dimensions, future people are likely to be better off than current people. Consider, for example, the improvements of recent decades in fields such as medicine, transportation, and telecommunications – we live longer, fly cheaper, and communicate more easily than even our recent forebears. And there is little reason to think this progress is about to come to a halt. Consider, too, that we have an international order that has not seen a world war for more than seventy years even though the interval between the First World War and the Second was only twenty-one years. Finally, consider the progress that has been made, over the last several decades and centuries, in curbing and eradicating unethical practices such as systematic discrimination on the basis of race, sex, sexual orientation, and (more recently) gender identity. Although moral progress is often unsteady – sometimes being "two steps forward, one step back" – it seems likely that future people will inherit an ethos of greater respect and inclusiveness than was experienced by previous generations. All things considered, then, the concept of intergenerational justice does not deliver the unequivocal verdict that the standard narrative often claims.

A second factor complicating the intergenerational-justice narrative is that climate change will not affect all future people the same, or even similarly. Even if it is correct to say that the climate is getting worse for humanity overall, this generalization masks two further facts that must not be overlooked. First, some people will benefit from climate change. Some residential real estate will become more valuable, and likewise for some commercial real estate, ranging from ports to tourism destinations. Some farmland will become capable of growing more valuable crops. And some people will have skills and expertise that will be in greater demand. Second, among people who are harmed by climate change, some people will be harmed a lot less than others. A corporate executive who owns a Miami Beach condominium that loses 20 percent of its resale value will probably have his lifestyle affected far less than a subsistence farmer in Indonesia who owns a small house that starts flooding from time to time. The condominium owner might have a greater dollar-value loss, but he can probably adapt to the changed circumstances with less effect on his well-being than the owner of the small house that soon might not even be habitable.

These facts not only show the inadequacy of the intergenerational-justice narrative, but also point the way toward a response to climate change that goes beyond just trying to lessen it altogether (though I do support that, as indicated by my discussion of a carbon tax). If we had perfect foresight and frictionless channels for reallocating resources, then it would make sense to simply identify and aid the people who will be harmed the most by climate change. Unfortunately, given the imperfect knowledge and institutions that we have, such a fine-grained approach is impractical. But there is a viable second-best approach, suggested by the example of the Miami Beach executive and the Indonesian farmer. We can expect that, to considerable extent, the poorer a person is, the more their well-being will be damaged by climate change. This means that providing aid to poor people is, in effect, a way of lessening the harms of climate change. And it is a much more targeted way of doing that than just trying to lessen climate change altogether.

In response to this argument, it is natural to think that if we were to just prevent climate change, then we wouldn't have to worry about finding exact or approximate ways of targeting the people adversely affected by it. However, climate change is basically unavoidable. Even if global GHG emissions were to drop sharply today, global average temperature would still continue to rise, because of the GHG accumulation already present in the atmosphere. Moreover, given current trends in major GHG-emitting countries such as the United States and China, the prospects for even modest reductions in global GHG emissions are dim. Thus, absent a technological miracle that could simply reverse past and future GHG emissions (such as massively scalable carbon-sequestration technology), we must acknowledge that significant climate change is going to occur, and that decisions we make now can put people in better or worse positions to cope with it.

Let me turn to two additional reasons favoring an increased focus on poverty reduction in this context. First, although we should be increasing our current GHG-reduction efforts, we must also keep in mind that each additional degree of cooling (or non-warming) will be more costly to achieve than the previous one: there is "low-hanging fruit" that we can grab relatively easily, so that we might achieve 1 degree of cooling pretty inexpensively, but the second 1-degree increment will be more costly to achieve, with the third 1-degree increment being even more costly, and so on. (For a homely analogy, consider that when trying to lose weight, losing a few pounds might be relatively easy, losing another few pounds is typically more difficult, losing another few pounds is even more difficult, and so on.) In other words, money spent on GHG-reduction efforts has decreasing marginal effectiveness, in terms of the quantity of the resulting GHG reductions and temperature reductions. Consequently, there will come a point at which we can more effectively lessen the harms of climate change not by plowing more money into preventing it, but by letting it happen – to some extent – and using our resources to help the most vulnerable people adapt to it, whether by buying a home on higher ground or paying higher prices for energy or food.

A second reason favoring increased poverty-reduction efforts is that greater wealth would not only make poor people more capable of adapting to climate change, but would also help them deflect the impact of other calamities they may face. In addition to the financial burdens that poor people already face, they may face, at any time, any of a Pandora's box of crises that could affect everyone in a certain region or the entire planet, such as fallout from nuclear weapons and dirty bombs (either in conventional warfare or terrorism), chemical and biological weapons such as weaponized smallpox, accidental epidemics such as the 1918–20 influenza that killed 50 million people, abrupt disruptions to the food supply, failures of utility distribution systems (e.g., water, electricity, or natural gas) due to accidental or malicious disruptions to their control systems, failures of other communications systems (personal phones, the global banking system, etc.), and non-climate-based natural disasters such as earthquakes, tsunamis, volcanic eruptions, and meteor strikes. For people confronting the consequences of any of these events, it probably will not matter very much whether the atmosphere is a couple of degrees warmer or cooler than it might have otherwise been. But it probably will matter very much whether they can afford to relocate, or pay for medical care, or repair their homes in some way, or just not default on their mortgages because of unemployment stemming from the economic disruption that many of the above events would cause. Money cannot buy everything, but it can buy things that people need in a wide range of life-altering circumstances – with climate change being, unfortunately, just one of them.

I mentioned at the beginning of this discussion that poverty reduction and climate change are often seen as two separate issues. On this widely held view, any resources committed to one cause are thereby denied to the other, just as any dollar donated to the local symphony is thereby denied to the local animal shelter. But as I have argued here, poverty reduction and climate change overlap to a considerable extent. Since poverty makes people more vulnerable to being harmed by climate change, the threat of climate change makes poverty reduction an even greater priority than it has long been recognized to be. Thus, poverty reduction, along with more commonly discussed policies such as a carbon tax, should be seen as urgently required by the growing threat of climate change.

Note

1 I would like to thank Dale Miller, my colleagues in the Philosophy Department at the University of Kansas, and the audience at a session of the Philosophy, Politics, and Economics Society Conference in New Orleans on March 16, 2018 for their helpful comments.

Bibliography

Cowen, Tyler. 1992. "Consequentialism Implies a Zero Rate of Intergenerational Discount." In *Justice between Age Groups and Generations*, edited by Peter Laslett and James S. Fishkin, 162–68. New Haven: Yale University Press.

——— and Derek Parfit. 1992. "Against the Social Discount Rate." In *Justice between Age Groups and Generations*, edited by Peter Laslett and James S. Fishkin, 144–61. New Haven: Yale University Press.

Gardiner, Stephen M. 2011. *A Perfect Moral Storm*. Oxford: Oxford University Press.

——— and David A. Weisbach. 2016. *Debating Climate Ethics*. Oxford: Oxford University Press.

Goodin, Robert E. 1995. *Utilitarianism as a Public Philosophy*. Cambridge: Cambridge University Press.

Griffin, Jr., Christopher L. 2012. "The Alaska Permanent Fund Dividend and Membership in the State's Political Community." *Alaska Law Review* 29, no. 1 (June): 79–92.

Jamieson, Dale. 2008. *Ethics and the Environment*. Cambridge: Cambridge University Press.

Kestenbaum, David. 2013. "Economists Have a One-Page Solution to Climate Change." National Public Radio, June 28, 2013. https://www.npr.org/sections/money/2013/06/28/196355493/economists-have-a-one-page-solution-to-climate-change.

Kestenbaum, David, Alex Blumberg, and Robert Smith. 2018. "Episode 472: The One-Page Plan to Fix Global Warming … Revisited." National Public Radio, July 18, 2018. https://www.npr.org/sections/money/2018/07/18/630267782/episode-472-the-one-page-plan-to-fix-global-warming-revisited.

Lenski, Shoshannah M., Gregory A. Keoleian, and Michael R. Moore. 2013. "An assessment of two environmental and economic benefits of 'Cash for Clunkers'." *Ecological Economics* 96 (December): 173–80.

Malakoff, David, Robert F. Service, and Warren Cornwall. 2016. "Trump targets key climate metric." *Science* 354, no. 6318 (December 16): 1364–65.

Nordhaus, William. 2013. *The Climate Casino: Risk, Uncertainty, and Economics for a Warming World*. New Haven: Yale University Press.

Schelling, Thomas C. 1997. "The Cost of Combating Global Warming: Facing the Tradeoffs." *Foreign Affairs* 76, no. 6 (November–December): 8–14.

Shafer-Landau, Russ. 2018. *The Fundamentals of Ethics*, 4th edition. Oxford: Oxford University Press.

Singer, Peter. 1972. "Famine, Affluence, and Morality." *Philosophy and Public Affairs* 1, no. 3 (Spring): 229–43.

———. 2016. *One World Now: The Ethics of Globalization*. New Haven: Yale University Press.

Wynes, Seth, and Kimberly A. Nicholas. 2017a. "The climate mitigation gap: education and government recommendations miss the most effective individual actions." *Environmental Research Letters* 12, no. 7 (July): 1–9. https://doi.org/10.1088/1748-9326/aa7541.

Wynes, Seth, and Kimberly A. Nicholas. 2017b. Supplementary materials for "The climate mitigation gap: education and government recommendations miss the most effective individual actions." http://iopscience.iop.org/article/10.1088/1748-9326/aa7541/data.

4

THE RULE-CONSEQUENTIALIST RESPONSE TO CLIMATE CHANGE

Dale E. Miller

Introduction

In this chapter I will consider how the moral theory known as rule consequentialism answers the question of what moral obligations we bear in the face of climate change. If you are already a committed rule consequentialist, or at least positively inclined toward the theory, then you should naturally be interested in what it says about how we morally must respond to this looming environmental crisis. If you are not already positively inclined toward rule consequentialism, then you should still be interested in how the theory answers this question, but for a different reason. You should be interested because what rule consequentialism has to say is sufficiently persuasive that it may lead you to see the theory in a more attractive light.

Rule Consequentialism: A Primer[1]

By "rule consequentialism" I mean "ideal-code" rule consequentialism, of the sort made familiar by Richard Brandt (1979) and Brad Hooker (2000a).[2] Rule consequentialism's moral standard, or criterion for morally evaluating actions, can be characterized in terms of two tenets:

1) The moral standing of actions—whether they are obligatory, right, or wrong—depends on whether they are required, permitted, or forbidden by an "authoritative" moral code or set of moral rules, and
2) The authoritative moral code is the "ideal code," the code whose general adoption would be "optimal" (or "optimific"), that is, would yield the best consequences.

Rule consequentialism is sometimes described as a variety of "indirect" consequentialism, in contrast with "direct" act consequentialism. This reflects the fact that while act consequentialists apply the consequentialist criterion of producing the best result directly to actions, judging an action to be wrong unless its consequences are at least as good as those of anything else the agent could do instead, rule consequentialists apply this criterion directly to moral codes and only indirectly to actions. However, "sophisticated" forms of act consequentialism are also sometimes described as indirect, which makes the terminology of "direct" and "indirect" consequentialism unhelpful. Sophisticated act consequentialism holds that people should generally decide what to do not by explicitly trying to calculate how they could produce the best consequences, but rather by following rules or employing some other "decision procedure." But while this may give sophisticated act consequentialism and rule consequentialism a surface similarity, the two views have very different moral standards and so are quite distinct. While a rule consequentialist will say that an action that conformed to the ideal code was right even if its consequences were sub-optimal, a sophisticated act consequentialist will maintain that an action with sub-optimal consequences was wrong even if it was chosen via the recommended decision procedure.

Varieties of Rule Consequentialism

Both of these tenets, and especially the second, can be interpreted in multiple ways, with different versions of rule consequentialism corresponding to these different interpretations. Strictly speaking, therefore, rule consequentialism is not one moral theory but rather a family of theories. It is worth touching on some of the differences between theories under the rule-consequentialist umbrella.

Consider first what it means to adopt a moral code. Adopting a code might be understood to mean complying with it, so that someone has adopted a code just if she never violates it. Most contemporary rule consequentialists understand what it means to adopt a moral code in terms of accepting it, however, which is a psychological rather than a behavioral notion. More specifically, many rule consequentialists equate adopting a moral code with "internalizing" it; that is, feeling compunction when one contemplates violating the code and guilt if one does so anyway (e.g., Brandt 1979, 286–305; Hooker 2000a, 32). Internalizing a code might be described as having one's conscience enforce it. Understanding adoption in terms of internalization has some important implications. One is that there is some limit on how large or complex the ideal code can be, because we are not psychologically capable of internalizing too many rules or rules that are too complicated. Another is that, because there are "teaching" costs involved in getting people to internalize rules, these costs must be considered in determining the content of the ideal code. Various factors might influence the magnitude of these costs. The number and length of the rules in question would likely be one. How much sacrifice the rules require might be another; instilling rules that require extreme sacrifices could necessitate draconian measures.[3]

The fact that understanding adoption in terms of internalization limits the length and complexity of the ideal code actually helps rule consequentialists deal with certain objections, but if this were the only rationale for this move then it might appear ad hoc. There is another rationale, however; namely that it fits well with an analysis of the concept of morally wrong action that apparently finds its first expression in the work of the nineteenth-century rule consequentialist John Stuart Mill. On Mill's analysis, what it is for an action to be wrong just is for it to be appropriate for someone who performs the action to experience guilt as a result (Mill 1969, 246).[4] If we take wrongness and guilt to be conceptually related in this way, then this suggests that to characterize a rule of conduct as a moral rule—as opposed to a rule of etiquette or other sort of rule—just is to say that it should be internalized. If Mill is right, then every theory of right and wrong can be viewed as an answer to the question of which rules people should internalize. The rule-consequentialist answer is that they should internalize the rules whose internalization would yield the best consequences. While most of the distinctions that I discuss involve more than one viable option, I will henceforth assume that understanding adoption in terms of compliance is a non-starter and that adoption should be equated with acceptance and more specifically internalization.

Next, consider what it means for a moral code to be "generally" adopted. At its most capacious, this phrase might be construed to refer to the code's being adopted by everyone, everywhere, "everywhen"—throughout time. Some rule consequentialists do construe generality this comprehensively (see, e.g., Parfit 2011, 398–403). Most, though, depart in one way or another from an understanding of generality that is so all-encompassing.

First, some rule consequentialists are what I will call "communal relativists," who say that the authoritative moral code for a given individual is the one whose general adoption by members of some particular community or group to which that individual belongs would have the best consequences. The group they have in mind is typically very broad, for example, the individual's "society" (e.g., Brandt 1979,179–81). As Hooker points out, however, once we start down this road we might characterize the relevant groups even more narrowly, all the way down to "groups" of one person each (2000a, 87). Hooker offers this as a reason not to take the first step; he rejects communal relativism and hence is a "communal universalist."

Second, some rule consequentialists are "temporal relativists," who say that the authoritative moral code at a given moment is the one whose general adoption throughout some particular period that includes that moment would have the best consequences. This opens the possibility that the ideal code might evolve over time. Different forms of temporal relativism might say that the authoritative code for those of us alive today is the one whose general adoption from this point in time forward would have the best consequences, regardless of what might have happened had that code been adopted in the past, or that it is the code whose adoption throughout the current epoch of history (however that is defined) would be optimal. A rule consequentialist who rejects temporal relativism can be called a "temporal universalist." This distinction has received less attention than it deserves, and it is frequently difficult to tell which side of it a particular rule consequentialist falls on.

Finally, even rule consequentialists who reject both communal and temporal relativism may deny that we should equate "general adoption" with "unanimous adoption," since if we do then the ideal code may not include any rules governing the treatment of people who have not internalized the code. Hooker, for instance, has proposed that the authoritative code is the one whose internalization by 90% of the population would yield the best consequences (2000a, 80–85).[5] Of course, one can combine any two or all three of these ways of restricting or limiting the scope of "general adoption."

The phrase "best consequences" is also open to different interpretations. Like act consequentialists, rule consequentialists tend to construe the notion of consequences very broadly, so that the consequences of a given moral code's being internalized comprise the entire "future history of the world" from that point forward. Consequentialists typically answer the question of what makes one state of the world better than another by saying that the goodness of a state of the world is a function of how much it contains of whatever has intrinsic (in the sense of non-instrumental) value; that is, whatever is valuable for its own sake and not only as a means to something else. Some consequentialists are welfarists, who maintain that only well-being or happiness has intrinsic value, while some are not. Of those who are, some are utilitarians, who hold that the same quantity of well-being is always equally valuable regardless of who enjoys it—and hence that the best state of the world is the one that contains the largest sum-total of well-being possible. Others are prioritarians, who hold that the value of a "unit" of well-being is higher when it is given to someone who is faring worse than to someone who is faring better. Some of the most important figures in the development of rule consequentialism, including Mill and Brandt, are specifically rule utilitarians (Mill 1969, 210; Brandt 1979, 286–305). Hooker, in contrast, tentatively endorses rule prioritarianism (2000a, 59–65).[6] Consequentialists who are not welfarists might still hold that only one sort of thing has intrinsic value, but deny that this is well-being. Others might say that several distinct sorts of things all possess intrinsic value, of which well-being might or might not be one.

Rule consequentialists who agree about what has intrinsic value may still disagree about how to take account of the fact that we cannot know with certainty how the future history of the world would unfold if different moral codes were generally adopted. "Objective" or "actual-consequence" rule consequentialism says that the ideal code is the one whose general adoption would in fact yield the best consequences. "Prospective" or "expected-consequence" rule consequentialism says that the ideal code is the one whose general adoption would have the highest expected value (Hooker 2000a, 72–5; Mason 2014). The expected value of a code's adoption is the average of the values of the different possible future histories that might result from its adoption, with each value weighted by the probability that it would result.

What Is the Ideal Code?

No rule consequentialist has produced a complete list of the rules that she takes to constitute the ideal code. This may make it seem that it would be impossible for us to apply or evaluate the theory. However, consider what a massive undertaking it would be to list all of the rules of what we might call contemporary Western "ordinary morality," the moral code that is generally taught and internalized today in Western countries. Spelling out the ideal code would be no less daunting of a task.

In fact, regardless of what specific version of the theory they embrace, most rule consequentialists seem to assume that contemporary ordinary morality is a reasonable first approximation of the ideal code. There is a plausible rationale for this assumption, which is that ordinary morality would naturally have evolved to promote and protect the things that we consider valuable for their own sakes, especially human well-being (Mill 1969, 207). (Some caveats: We should expect ordinary morality to have evolved in ways that are biased toward the interests of those who enjoyed more social power, e.g., men. And there is admittedly a Eurocentric bias in taking Western morality specifically as a starting point, at least for rule utilitarians who purport to seek a code that is authoritative globally.)

The standard rule-consequentialist view, then, is that like ordinary morality the ideal code would comprise a plurality of rules that forbid or require broad categories of actions. And this seems to be correct. While J. J. C. Smart suggests that the ideal code might comprise a single rule, one that requires us to do the optimal action at every opportunity (Smart 1973, 11–12), there are good reasons to believe that it would not be optimal for a moral code comprising only this one rule to be generally internalized. Consider, for instance, how difficult it would be to predict how a person whose moral code comprises just this one rule would behave, given that this would require knowing her beliefs about all the consequences of all the things that she might do; this would make it very hard for the inhabitants of a world in which everyone internalized only this one rule to cooperate or otherwise coordinate their behavior (Harsanyi 1998, 292–3; Hodgson 1967, 38–50).[7]

Like ordinary morality, the rule-consequentialist ideal code might contain rules that are formulated at different levels of specificity and generality. It might also contain rules that are formulated at different levels of concreteness or abstractness. Including abstract rules—principles—or very general rules in a moral code could represent greater efficiency, since it could reduce the number of rules the code contains. Abstract principles in particular could offer people guidance in wide ranges of situations in which the choices that they face are very different. Even a principle as abstract as "Always treat people fairly." might be part of the ideal code, as long as we are capable of internalizing it. However, from a consequentialist standpoint there are disadvantages to rules that are too general; the general adoption of a rule that prohibits all lying, without exception, could hardly be expected to yield better consequences than the adoption of rules that treat different varieties of lies differently. Moreover, abstract principles must be interpreted to be applied, which creates the potential for both misinterpretation and—because it can be difficult to predict how someone else will interpret an abstract principle—difficulty in coordinating behavior. So there are countervailing costs and benefits that influence what sorts of rules will make up the ideal code. The optimal solution will likely combine rules with different degrees of generality and abstractness. In some cases, the ideal code might contain both abstract principles and more concrete rules that amount to applications of those principles; this would allow for the benefits of abstractness to be enjoyed while still ensuring that the same interpretations of the principles were widely shared when the benefits of coordination would be most valuable.

Just as the rules that constitute the ideal code may vary in their levels of generality and abstractness, so too might they vary in what we might call their "absoluteness." A maximally absolute rule admits of no exceptions. If it uniformly forbids some category of actions, then actions of that sort are always wrong, all things considered. There are at least two ways in which rules might fall short of this level of absoluteness, however. First, they might incorporate exception clauses. For instance, a rule against lying might say that it is wrong to lie *except when* a trivial lie is necessary to spare someone's feelings, or when a malefactor is seeking information that will be used to do something morally repugnant, or when .... Second, they might sometimes be outweighed or defeated by other rules with which they conflict. In some instances, one rule might always take priority over another, which makes the former "lexically prior" to the latter (Rawls 1999, 37). In other instances, which of two conflicting rules takes precedence might depend on how severe of a violation of one it would be to obey the other in a given case (Brandt 1965, 130–4; Hooker 2000a, 88–92; cf. Eggleston 2007). (When a rule falls short of being maximally absolute it may sometimes be arbitrary whether we think of it as incorporating exception clauses or as occasionally being outweighed by other rules. Talking about rules is really a convenient way to talk about sets of dispositions to experience attitudes such as compunction and guilt. There might be multiple equivalent ways to "translate" an individual's dispositions to experience these attitudes into explicit rules.)

In addition to rules that forbid or require certain actions, the ideal code might also contain rules that grant agents certain permissions or, to use Samuel Scheffler's term, "prerogatives" (Scheffler 1994); these might, for instance, be permissions for agents to promote their own good in certain circumstances. Sometimes, these "permission-conferring" rules might outweigh "obligation-imposing" rules. Suppose that a professor has promised to meet a student in her office at a specified time, but that due to circumstances beyond her control she can now do this only at great personal cost, for example, by forgoing the opportunity to redeem a winning lottery ticket. In this case, we might plausibly think that a permission-conferring rule allowing her to promote her own financial well-being takes priority over the rule requiring promises to be kept. By combining obligation-imposing and permission-conferring rules, the ideal code might forbid certain sorts of actions unless they would make an especially large contribution to the agent's well-being.

Arguments for Rule Consequentialism

Rule consequentialists have offered a variety of arguments in favor of the theory. Hooker has very influentially argued for rule consequentialism on the grounds that it resembles our pre-theoretical "ordinary morality" at least as closely as do any other philosophical theories of ethics (Hooker 2000a, 4–30). He presents this argument as an application of the "reflective equilibrium" methodology (Hooker 2000b). To state the point more carefully, Hooker favors rule consequentialism because he believes that it fits widely shared convictions about the rightness or wrongness of actions, convictions that are sometimes called "moral intuitions" or "considered moral judgments," at least as closely as any other moral theory. Moreover, he believes that it enjoys certain formal advantages over the only rival theory—the moral pluralism associated with W. D. Ross (2002)—with which it ties in this respect. From this he concludes that subscribing to rule consequentialism is the best way to bring our moral beliefs into mutual coherence. One possible weakness in Hooker's methodology is that it is extremely conservative, in the sense of taking most of what is commonly believed about our moral obligations for granted rather than subjecting it to critical scrutiny (D. Miller 2000; cf. Hooker 2000b, 229–31). Another is that the moral intuitions to which Hooker appeals, namely those of educated contemporary Westerners, are highly idiosyncratic (Haidt 2012, 95–111). Yet he concludes not only that rule consequentialism is universally justified but even that the same ideal code is authoritative everywhere, everywhen.

Hooker's argument is not premised on an "overarching commitment to maximize the good." Other arguments for rule consequentialism do rest on the claim that we ought to be committed to producing the best outcomes possible. These arguments try to show that this commitment further commits us to rule consequentialism. For example, the economist John Harsanyi argues for rule utilitarianism on "contractualist" grounds (e.g., Harsanyi 1998).[8] He asserts that the authoritative moral code is the one that would be agreed upon by parties in the sort of impartial choice situation that Rawls (1999) famously calls the "original position," and that this is the code whose general adoption would result in the highest level of social utility possible—the ideal code. Harsanyi therefore concludes that we should be rule utilitarians.

As it stands, Harsanyi's argument fails. Recall that sophisticated act consequentialists maintain that people should generally decide what to do not by trying to explicitly calculate which of the actions open to them would produce the best consequences but by following some other decision procedure. This decision procedure might involve internalizing a set of rules (Hare 1981, 25–64). So act and rule utilitarianism, say, might largely agree about what moral code should be generally adopted, despite the fact that they have distinct moral standards (Levy 2000). In that case, knowing that parties in the original position would agree that this moral code should be generally adopted does not help us choose between these theories.

However, it may be possible to strengthen Harsanyi's argument. Recall the Millian understanding of what it is for an action to be wrong that was described earlier, according to which the concept of morally wrong action can be analyzed in terms of the appropriateness of guilt. Given this analysis, if a moral theory says that we ought to internalize a particular moral code, then that theory entails that actions contrary to this code are wrong. If the theory also denies that those actions are wrong, then it is inconsistent. And this would mean that sophisticated act consequentialism is inconsistent, in virtue of entailing both that actions that violate rules like those in the rule-utilitarian ideal code are always wrong and that actions that violate those rules are not wrong whenever they would have the best consequences. Since surely our moral theory ought to be internally consistent, someone following Harsanyi's approach to moral theory selection now has reason to favor rule over act utilitarianism. And this consideration has implications that extend beyond Harsanyi's specific argument; it raises a general worry about sophisticated act consequentialism. It may give us reason to see any argument that entails that the moral code that should be generally adopted is the one whose general adoption would be optimal as an argument specifically for rule consequentialism.

The "Rule Worship" Objection to Rule Consequentialism

Numerous objections have been raised against consequentialism generally and rule consequentialism specifically. I will discuss just one of these here, the "incoherence" or "rule worship" objection. Considerations of space will still prevent me from delving too deeply into the issues it raises or possible responses.

The incoherence objection is an argument against rule consequentialism that is pressed most frequently (albeit not exclusively) by act consequentialists. Smart gives a general sense of the objection when he asks why, since "the rule-utilitarian presumably advocates his principle because he cares about human happiness," he should "advocate abiding by a rule when he knows it will not in the present case be most beneficial to abide by it?" (Smart 1973, 10; see also Smart 1956, 353). The incoherence objection is best understood as an argument for the claim that rule consequentialists are committed to inconsistent propositions about our reasons for action; namely that we always have the most reason on balance to do the optimal action and that we have conclusive reason not to do the optimal action when it is forbidden by the ideal code (D. Miller 2011, 95–97). The critic takes the rule consequentialist to be committed to the first of these propositions by any argument for why it is the ideal code that is authoritative, on the assumption that any argument for this conclusion will be premised on our always having reason to prefer optimal options, and to the second by the fact that we always have conclusive reason to act morally. One way for a rule consequentialist to rebut this objection is to show that she is not in fact committed to one or the other of these propositions. Hooker tries to do this by giving an argument for rule consequentialism that is not premised on an overarching commitment to maximize the good. Another possibility would be to reject the critic's assumption that we always have conclusive reason to act morally.

Rule Consequentialism and Climate Change: The 50,000 Foot View

In this section I will describe in fairly abstract terms how rule consequentialists might approach the issue of climate change, before venturing some more detailed and concrete prescriptions in the next.

There is no serious debate about whether, if current trends in emissions of greenhouse gases (GHGs) by humans continue, the Earth will experience climate change and concomitant ills—e.g., sea-level rise, extreme weather events, and the acidification of the oceans—that will result in untold human suffering. Non-human animals will suffer, too, and many species will go extinct. Humanity's own survival is by no means assured. Call this "the future toward which we are heading" (FTWWAH). If we reach FTWWAH, it will be as the result of the combination of many individual actions by many individual agents over time, where each of the individual actions is quite mundane: driving a car, ordering a hamburger, buying an airline ticket, turning on the lights. The performance or non-performance of any one of these actions, considered in isolation, may well make no difference to how much suffering results from climate change.

If it is still within our power to avoid FTWWAH to a meaningful extent—if not to avert climate change entirely, because it is too late for that, then at least to avoid the worst parts of it—then are the necessary changes to our behavior morally required? I find the thesis that they are strongly compelling. To answer otherwise, and so to believe that it could be morally permissible for human beings to bring about a disaster of this epic scale for humanity (and not only for humanity), is to see morality as objectionably disconnected from human concerns—as, in an important sense, not "for us." At least this is true unless the only way to avoid FTWWAH is to do something that would itself be absolutely morally impermissible, but prima facie that does not appear to be the case here. No one is proposing that we fight climate change with genocide.

Rule consequentialism seems to be uniquely well positioned to guarantee the truth of the thesis that we are morally required to do what we must if we are to avoid FTWWAH. If a general practice of freely engaging in some activity that produces GHG emissions will have significant negative consequences, then this is a strong reason to believe that the ideal code contains a rule forbidding that practice and restricting the circumstances in which the activity in question is permissible. Actions contrary to that rule will be wrong regardless of their individual consequences.[9] I do not want to prejudge the ability of rival moral theories to guarantee the truth of this thesis, but proponents of these theories will at least need to do more work to show how they do so. Consider, for instance, act consequentialists. Like rule consequentialists, act consequentialists can start by envisioning some very bad state of affairs that might arise in the future and then "work backwards," judging the conduct that would produce that state of affairs to be wrong. But act consequentialists can only do this on an action-by-action basis, so if taken singly mundane actions like taking a short drive make no difference to how much future suffering occurs then act consequentialists may struggle to explain why they are wrong. They may, in other words, have a hard time explaining how they avoid Derek Parfit's "Second Mistake in Moral Mathematics" (1984, 70–73), the mistake of thinking that if an action is wrong because of "effects" or consequences it must be wrong because of its own particular effects.[10]

At this point, however, many of the distinctions I drew in the last section between different versions of rule consequentialism come into play. To begin with, any consequentialist who is a welfarist, or even any consequentialist who makes the weaker assumption that well-being has considerable intrinsic value, must regard FTWWAH as catastrophic. But we can in principle imagine a rule consequentialist with some idiosyncratic ideas about what has intrinsic value according to which FTWWAH would not be a bad state of affairs (e.g., one who believes that the only thing desirable as an end is heat). Happily, since virtually all consequentialists do give well-being a prominent place in their "theories of the good," we can set this possibility aside.

In contrast, there may be no obvious answer to the question of whether rule consequentialists should be temporal relativists, and if so of what sort. Yet what position a rule consequentialist takes on this question may make a very significant difference to what she does or can say about climate change. There are likely some moderate rates of GHG emissions such that, if humans had kept to these rates throughout history, they could have enjoyed many of the benefits of burning fossil fuels without ever needing to worry about severe climate change or other significant negative consequences. A temporal universalist's ideal code might contain rules that require those rates not to be exceeded. Given that past emissions of GHGs far exceeded these moderate rates, however, for us to begin to conform to these rules today would not suffice for us to avoid FTWWAH. So a temporally universalist version of rule consequentialism might not guarantee that the measures that we would need to take today to avert FTWWAH are morally obligatory. In contrast, a temporally relativist version of rule consequentialism that characterizes the ideal code as the one whose general adoption from this point forward would be optimal will presumably require whatever measures we would need to take starting now to escape FTWWAH.

The distinction between versions of rule consequentialism that do and do not reflect communal relativism may be important here, too. One of the most important questions in climate ethics is that of how the burden of avoiding FTWWAH should be divided between economically "developed" countries and those that are still developing. The rule consequentialist answer to that question may differ depending on whether or not we assume that there is a single ideal code whose general acceptance globally would be optimal. Here is one way in which this might be true. Suppose that there are distinct "developed world" and "developing world" ideal codes, the content of which depends on which rules' general acceptance would be optimal in their respective spheres. When rule consequentialists ask which moral code would be ideal, they imagine the world's being exactly as it is except for what moral code is adopted; they ask, that is, which code's adoption would have the best consequences in the "real world." If there is a separate ideal code for the developing world, then consistency seems to require the rule consequentialist to identify this with the code whose general adoption there would be optimal *given the developed world as it actually is*, not as it would be if it adopted its own ideal code. Sadly, given American recalcitrance the probability that the developed world will do enough to stave off FTWWAH may be so low, realistically speaking, that the ideal code for the developing world might place few if any restrictions on GHG emissions. In other words, the future may be so tragic, regardless of what the developing world does, that those poorer nations might as well enjoy the benefits of the profligate use of fossil fuels now. Even if this would make the future slightly worse, by causing the world to reach FTWWAH slightly sooner, the benefits to people alive now may outweigh whatever marginal future suffering would be experienced. Humanity can only go extinct once.

The distinction between prioritarian and utilitarian versions of rule consequentialism also bears on the question of how the burdens of reducing GHG emissions should be divided. Rule utilitarianism may well entail that the developing world should have to shoulder comparatively little of the burden of avoiding FTWWAH; the principle of "diminishing marginal utility" suggests that from the utilitarian standpoint it is better for those with more wealth to contribute more. But if rule utilitarianism entails that the developing world should shoulder very little of this cost, rule prioritarianism will entail that it should shoulder even less. A rule prioritarian will take the subtraction of a unit of well-being from a person who is relatively worse off to do more to worsen the world than the subtraction of the same amount of well-being from a person who is faring better. In consequence, she will believe that fewer sacrifices should be demanded of those whose levels of well-being are already lower. And while I have framed this point specifically in terms of prioritarianism and utilitarianism, it can be generalized to any rule consequentialists who believe that well-being has intrinsic value, whether or not they are strict welfarists.

So far, I hope, nothing has emerged in my discussion to give us reason to worry that rule consequentialism is somehow incapable of offering a plausible account of moral obligations in the face of climate change. However, the utilitarian philosopher and environmental ethicist Dale Jamieson has recently argued that even committed consequentialists ought to see rule consequentialism, in company with act consequentialism, as inadequate to this task. I will close this section by considering his objection.

Act utilitarianism, Jamieson argues, is beset by the problem of "contingency" (2007, 167–68). This is the challenge for act utilitarians and act consequentialists generally to which I alluded previously, that of explaining why the individual mundane actions responsible for producing climate change are wrong. But Jamieson says that the same fundamental problem faces rule consequentialists, in company with "motive consequentialists" and others:

> Ultimately, the most important problem with act-utilitarianism is also a problem with indirect views that focus on motives, rules, or whatever. All of these accounts are 'local', in that they privilege some particular 'level' at which we should evaluate the consequences of actions that are open to us. Rather than adopting any such local view, we should be 'global' utilitarians and focus on whatever level of evaluation in a particular situation is conducive to bringing about the best state of affairs.
>
> *Jamieson 2007, 170; see also Driver 2014*

Jamieson asserts that the "level" upon which we should focus when it comes to actions that could make some difference to climate change is that of "non-calculative generators of behavior: character traits, dispositions, emotions and what I shall call 'virtues'" (2007, 167). He concludes that in response to the threat of global warming we ought to embrace a "virtue ethics" that incorporates certain "green virtues" including humility, temperance, and mindfulness (2007, 181–82).[11]

Jamieson's critique cuts no ice as an objection to ideal-code rule consequentialism. The rules on which this theory focuses are not mere verbal formulas but rather internalized rules; that is, dispositions to experience attitudes (or feelings or emotions) such as compunction and guilt. They come with motivations attached, in other words, and so are as much "non-calculative generators of behavior" as virtues. Whether they would be better or worse generators of behavior than virtues, from a utilitarian perspective, is an important question but not one that Jamieson takes up. Of course, a rule consequentialist need not deny that it is desirable for people to have certain virtues, just as Jamieson need not deny that it is desirable for them to internalize certain rules. As I have characterized rule consequentialism, though, in determining which rules constitute the ideal code we must not imagine that people are any more virtuous than they actually are (except when their becoming more virtuous would be a consequence of their internalizing the rules in question). There is room in conceptual space for a "virtue-plus-rule" consequentialism which asks which combination of virtues and internalized rules it would be best for people generally to possess. The rules that constitute part of this ideal combination might be rather different from the rules of the "straight" rule-consequentialist ideal code, just as the virtues might be somewhat different from those of a "straight" virtue consequentialism. The optimal division of labor between rules and virtues would have to be determined. But while exploring this view would be a worthwhile project, what matters at present is that Jamieson has given rule consequentialists no cause for worry.

What Rule Consequentialism Requires

In order to work out in more detail what rule consequentialism requires from us in the face of climate change, I will need to settle on a specific version of the theory. Moving forward, therefore, I will assume without argument that the most plausible form of rule consequentialism is a prospective rule utilitarianism, one that is temporally relativist in the sense of identifying the ideal code with the set of rules whose general adoption would have the best consequences from today forward and communally relativist to the extent of holding that the authoritative moral code for me is the one whose general adoption throughout what is sometimes called the "developed West" would be optimal. What interests me at present is what obligations the ideal code of the developed West imposes on those of us in this part of the world.

It is utterly clear that if FTWWAH is to be averted then the citizens of the developed West will need to dramatically reduce their carbon footprints. This rule-utilitarian theory's ideal code might include various abstract principles that could require these reductions, such as principles forbidding harmful conduct or requiring the preservation of non-human nature. However, because it is too uncertain how people who had internalized these principles might interpret them, and because it is too important that significant reductions in emissions be achieved to leave this to chance, the theory's ideal code will very probably contain an obligation-imposing rule prohibiting activities that involve the release of greenhouse gases as such.[12] While the theory will have to allow for this rule to be outweighed by other moral rules sometimes, including some permission-conferring rule or rules allowing people to pursue their own goods, the former rule will have to be weighty enough relative to the latter that the theory will entail that many of the mundane activities that citizens of the developed West now perform regularly—even daily or hourly—are forbidden. It is not possible for me to catalog all of the activities that the theory would likely prohibit; at most I can offer a few examples.

First, the rule-utilitarian theory that I have sketched will very likely forbid the consumption of ruminants like cattle and sheep. Their digestive process involves fermentation that releases methane. Methane is a potent GHG; although it does not remain in the atmosphere nearly as long as CO_2—around 10 years, as opposed to 200—while there it traps many times more heat. One recent commentary observes that "Globally, ruminants contribute 11.6% … of all greenhouse gas emissions from anthropogenic sources" (Ripple et al. 2014, 2). Estimates of methane emissions from ruminative digestion in the United States alone range from 6.2 million metric tons (Hristov et al. 2014) to nearly 17 million metric tons (S. Miller et al. 2013) annually. Even this lower figure would mean that American cows and other ruminants inject the equivalent of 155 million metric tons of CO_2 into the air each year, as much as burning 17.4 billion gallons of gasoline (Environmental Protection Agency 2018). And it is not only a matter of bovine belching: manure gives off even more methane, and if we no longer consumed these animals then at least some of the land on which they are raised could be reforested, which would remove some CO_2 from the atmosphere. Arguably, rule utilitarianism might prohibit the consumption of ruminants on grounds of animal welfare alone, without considerations of GHG emissions even being brought into the calculation. It is also possible, however, that it would permit at least the occasional consumption of their flesh if the animals were humanely raised and slaughtered. However, when considerations of GHG emissions are considered, they almost certainly tip the balance in favor of taking these animals out of our diet.

Second, the theory will require significant reductions in travel in vehicles powered by fossil fuels. The transportation sector is responsible for the largest share of GHG emissions in the United States, recently surpassing power generation in this respect (U.S. Energy Information Administration 2017). Emissions from transportation are also increasing in the European Union (European Environmental Agency 2017). Making these reductions will mean many fewer trips by automobile, obviously, and more trips by public transit—or, even better, by bicycle and walking. They will also mean taking many fewer flights. Not only do airplanes emit large quantities of GHGs and aerosols that contribute to warming, they do so in the upper atmosphere, which magnifies their impact (Aschwanden 2015). According to one online calculator, in two trips from Virginia to present versions of this chapter at conferences in New Orleans and Germany, I was responsible for the equivalent of over 4 metric tons of CO_2 (Atmosfair, no date). Since much of air travel is for business, individual travelers may be able to justify trips that are genuinely requirements of their jobs by appeal to the permission-conferring rule or rules that allow them some scope to promote their own good. However, this only shifts the obligation to businesses to find ways to reduce the need for their employees to travel. Even if academics can justify *some* conference travel as long as this is a professional expectation, for instance, universities and conference organizers are obligated to find alternatives to traditional conferences that do not require travel. If universities committed to making smaller reimbursements for travel but larger expenditures on telepresence technology, alternatives would emerge.

In order to ensure that individuals comply with this rule, politicians would be obligated to adopt measures that would supply economic incentives for reducing fossil fuel use, such as a carbon tax or cap and trade scheme. Such legal mechanisms might seem to be redundant, since if the theory's ideal code were generally adopted then people would reduce their emissions even in their absence. However, there would still be a point to deterring free riders via the law. I have defined the ideal code as the one whose internalization by the great majority of citizens would be optimal, so the ideal code would require us to find ways to deal with those in the minority. Moreover, even people who have internalized a moral code backslide sometimes.

So far I have been discussing a rule that imposes a "negative" obligation on us, a "thou shalt not." But when we ask just how stringent this rule is, how far it would require us to reduce our carbon footprints, we discover that this negative obligation does not go far enough. One of the distinctive features of rule consequentialism, at least compared with act consequentialism, is that it typically says that what we are required to sacrifice in the interest of solving large chronic problems is limited to whatever level of contribution would be sufficient were more or less everyone else to contribute the same, even if everyone else is not contributing that much. Hooker, for instance, proposes that no one would be obligated to give more than 10% of their annual income to fight global poverty, since if everyone were willing to give at most that much the problem could be solved (Hooker 2000a, 159–74).

Thus one way to think about how far the theory would require individuals in the developed West to reduce their carbon footprints would be to ask how much CO_2 or the equivalent an individual in this part of the world could be responsible for emitting, consistent with the avoidance of FTWWAH. In doing this calculation, it is necessary to make realistic assumptions about GHG emissions in the rest of the world. Complicating matters is the fact that I have only vaguely characterized FTWWAH, so that I am in no position to say with any precision what would count as avoiding it. For the sake of simplicity, let us operationalize this by saying that avoiding FTWWAH means staying below 2°C of warming relative to pre-industrial temperatures. Our remaining carbon budget for staying below this level of warming, with 66% certainty, has been estimated to be in the neighborhood of 700 billion metric tons of CO_2 (CarbonBrief 2017). Projections for future emissions outside of the developed West rise from about 24 or 25 billion metric tons of CO_2 equivalent annually now to about 31 billion tons in 2040 (U.S. Energy Information Administration 2016).[13] So by 2040 about 86% of the carbon budget would have been expended even if net emissions in the developed West dropped to zero today. Suppose optimistically that after 2040 emissions outside of the developed West remain at projected 2040 levels, rather than continuing to rise, and that the population of the developed West does not significantly change. In that case, to remain within our carbon budget by the year 2100 would require each citizen of the developed West to produce about *-27* metric tons of CO_2 or the equivalent each year, starting now. In other words, they would be required to purchase or produce carbon offsets equal to 27 tons more than the emissions for which they are responsible. For the average American, that would mean offsets of 43 tons per year (World Bank, no date). Thus this back-of-the-envelope calculation suggests that the rule-utilitarian theory I have sketched would impose a positive obligation to contribute to reducing GHGs—a "thou shalt" as opposed to a "thou shalt not."

One can purchase this quantity of carbon offsets for about the price of a typical American cable television bill, for example from groups who will plant enough trees to absorb the requisite amount of carbon from the atmosphere. But even if these are legitimate offsets, which is often uncertain (van Kooten 2017), they are low-hanging fruit. Everyone cannot buy this level of offsets so cheaply; there is only room to plant so many trees. For everyone or nearly everyone in the developed West to achieve a negative carbon footprint of the requisite size would require a very large transfer of wealth and/or technology to countries in the developing world to enable and incentivize them to limit their emissions well below current projections. This will surely require political action by Western governments. And so the positive obligation imposed by this rule-utilitarian theory's ideal code will extend to an obligation on the part of politicians to make the necessary commitments and of citizens to demand that they do so—and to pay the taxes that would be necessary to fund such measures. These commitments would presumably take the form of international "polluter pays" treaties on which the lion's share of the costs of mitigation, amelioration, and adaptation would be paid by the developed world. I do not know how to put a price tag on these costs, but paying them might easily exceed the level of expenditure that Hooker calls for in fighting poverty. The expense will of course depend in no small part on how far the developed West is able to reduce its own carbon emissions. Individually or collectively, the less that you emit, the less that you need to spend on offsets.

So far I have not addressed the question of what specific rules this rule-utilitarian theory's ideal code would contain that would be responsible for imposing this positive obligation on us. Abstract principles of beneficence might play a role, along with—in view of the West's massively outsized past emissions—a principle of restitutive justice that requires those who are most responsible for creating a problem to be responsible for most of the costs associated with addressing it. The latter principle has a clear utilitarian justification in the incentive it gives agents not to create problems. Principles requiring the preservation and restoration of non-human nature may also come into play. But once again, the theory's ideal code might have to include a more concrete rule that is much more specific to greenhouse gases, one that requires us to make a positive contribution toward restoring these gases to "natural" levels, to ensure that no one is able to convince themselves that they have satisfied morality's demands without contributing their share to this effort.

Conclusion

I will conclude my discussion by considering an objection that might be raised to my attempt to apply rule-consequentialist reasoning to moral challenges raised by climate change in the preceding section. Doing so will allow some parting comments on the differences between rule consequentialism and its chief "intramural" rival, act consequentialism.

A critic might claim that the combination of obligation-imposing and permission-conferring rules that I described in the last section looks like covert prospective act utilitarianism: follow the rule against emitting GHGs except when it would maximize expected utility to depart from it. But this is not the case, for at least two reasons. First, the rule-utilitarian theory that I am considering would not necessarily allow the use of fossil fuels whenever the expected utility of doing so would even slightly exceed that of forgoing their use. Compare: Act utilitarianism allows promises to be broken whenever the expected utility of promise breaking would be even a tiny bit greater than that of promise keeping. The rule-utilitarian ideal code would presumably allow promises to be broken when *enough* well-being is at stake, but only some significant amount would count as enough (Hooker 2000a, 98–99; Hodgson 1967, 40–42). Thus rule utilitarianism sets a "higher threshold" for permissible promise breaking than does act utilitarianism. This rule-utilitarian theory might similarly impose a higher threshold for the permissible use of fossil fuels.

Second, the two theories would reach very different verdicts in some cases. Suppose that Jill would benefit so much from using a significant quantity of fossil fuels that it would maximize expected utility for her to do so, even though she is the only one who would benefit. Perhaps she is considering taking a transatlantic flight, both for a much-needed vacation and so that she can do research for a personal genealogical project in which she is very invested. Prospective act utilitarianism would say that Jill is actually *required* to make the trip; she is morally obligated to do so. In contrast, the rule-utilitarian theory that I have sketched would say at most that it is *permissible* for her to take the flight. It might even say that it is morally commendable—supererogatory—for her to sacrifice her own good and forgo it.

Cases in which the two theories reach different verdicts about the permissibility of using of fossil fuels may be more frequent than is immediately apparent. If the available evidence points to most Westerners being unwilling to change their behavior enough to avert FTWWAH, then prospective act utilitarianism may actually require you to use fossil fuels quite freely yourself, since this would benefit you at least slightly without making the future appreciably worse. As long as FTWWAH can be avoided through the collective actions of your "moral community," in contrast, the rule-utilitarian theory will require you to obey the rule whose general adoption would be optimal, even if it is in fact clearly not being generally adopted and your following it would not be optimal. Admittedly, this shows that rule consequentialism sometimes has a quixotic quality that may seem unbecoming in a consequentialist theory. Why should a consequentialist favor expending time and treasure in what the evidence suggests is a lost cause?

Yet it is important to remember that rule consequentialism is a theory of what morality requires from us, not merely a strategy for producing the best consequences. There is more to justifying a moral theory than showing that it fits well with things that we already believe about morality, as I noted in my critique of Hooker's argument for rule consequentialism. Nevertheless, moral common sense supports the notion that sometimes morality requires acting as others should act, not as they do act. The Golden Rule tells us to treat others as we wish to be treated, not as they treat us.[14]

Notes

1 For a fuller treatment of the issues discussed in this section, see D. Miller (2014).

2 The phrase "ideal code" originally comes from Brandt (1979, 297). Ideal-code rule consequentialism can be contrasted with "actual-rule" rule consequentialism, which I do not discuss here but which says that rules must actually be adopted in order for them to be authoritative (R. Miller, 2009). Brandt is also due some credit for the term "rule consequentialism" itself, having coined "rule utilitarianism" and "act utilitarianism" (1959, 253, 380).

3 For discussion of these and related points see Hooker (2000a, 75–80).

4 This (controversial) reading of Mill owes much to David Lyons's pioneering work. For a thorough discussion of Mill's analysis of moral wrongness and his rule consequentialism see D. Miller (2010, 79–110).

5 A large literature has recently sprung up around the question of what level of social acceptance we should consider when trying to determine which moral code's acceptance would produce the best consequences (see, e.g., Smith 2010).

6 Hooker suggests that virtue also has intrinsic value, which strictly speaking makes him a pluralist about "the good," but he denies that its value should be considered in determining which rules constitute the ideal code (Hooker 2000a, 33–7; see also D. Miller 2013).

7 While Smart and Harsanyi both focus on rule utilitarianism, the point generalizes to rule consequentialism.

8 Parfit (2011, 375–420) also argues for rule consequentialism on contractualist grounds.

9 Aiste Seibokaite discusses rule consequentialism as a solution to the collective action problem presented by climate change (2015).

10 This is a correction of Parfit's own formulation (see Jackson 1997, 52 n. 3).

11 Jamieson's distinction between virtues and character traits, dispositions, and emotions is puzzling, since he approvingly quotes Julia Driver's definition of a moral virtue as "a character trait that systematically produces or gives rise to the good," and adds that this definition should be supplemented by some mention of the emotions (Jamieson 2007, 171).

12 The ideal code seems to be more likely to contain a general rule that prohibits the release of greenhouse gases as such than more specific rules that prohibit specific activities like driving or flying. After all, it is possible to drive an electric car powered by

renewable energy, and while it is not yet possible to fly in an electric airliner it very well may be in the not-too-distant future.

13 More specifically, these estimates pertain to countries outside of the Organisation for Economic Co-operation and Development (OECD) in combination with OECD countries in Asia.

14 I am grateful to Ben Eggleston and audiences at the 2018 meetings of the Philosophy, Politics, and Economics Society and the International Society for Utilitarian Studies for their helpful suggestions.

Bibliography

Aschwanden, Christie. 2015. "Every Time You Fly, You Trash the Planet—And There's No Easy Fix." *FiveThirtyEight*. https://fivethirtyeight.com/features/every-time-you-fly-you-trash-the-planet-and-theres-no-easy-fix/, accessed 24 March 2018.

Atmosfair. no date. https://www.atmosfair.de/en/, accessed 24 March 2018.

Brandt, Richard B. 1959. *Ethical Theory: The Problems of Normative and Critical Ethics*. Englewood Cliffs: Prentice Hall.

———. 1965. "Toward a Credible Form of Utilitarianism." In *Morality and the Language of Conduct*, edited by Hector-Neri Castañeda and George Nakhnikian. Detroit: Wayne State University Press, 107–43.

———. 1979. *A Theory of the Good and the Right*. Oxford: Clarendon Press.

CarbonBrief. 2017. "Analysis: Just Four Years Left of the 1.5C Carbon Budget." https://www.carbonbrief.org/analysis-four-years-left-one-point-five-carbon-budget, accessed 13 March 2018.

Driver, Julia. 2014. "Global Utilitarianism." In *The Cambridge Companion to Utilitarianism*, edited by Ben Eggleston and Dale E. Miller, 166–76. Cambridge: Cambridge University Press.

Eggleston, Ben. 2007. "Conflicts of Rules in Hooker's Utilitarianism." *Canadian Journal of Philosophy* 37 (3): 329–49.

Environmental Protection Agency. 2018. "Green House Gas Equivalencies Calculator." https://www.epa.gov/energy/greenhouse-gas-equivalencies-calculator, accessed 11 March 2018.

European Environmental Agency. 2017. "Analysis of Key Trends and Drivers in Greenhouse Gas Emissions in the EU Between 1990 and 2015." https://www.eea.europa.eu/publications/analysis-of-key-trends-and/, accessed 31 March 2018.

Haidt, Jonathan, 20102. *The Righteous Mind: Why Good People are Divided by Politics and Religion*. New York: Pantheon.

Hare, R. M. 1981. *Moral Thinking: Its Levels, Method and Point*. Oxford: Clarendon Press.

Harsanyi, John. 1998. "A Preference-Based Theory of Well-Being and a Rule-Utilitarian Theory of Morality." *Game Theory, Experience, Rationality: Vienna Circle Institute Yearbook*, Vol. 5, 285–300. Dordrecht: Kluwer.

Hodgson, D. H. 1967. *Consequences of Utilitarianism: A Study in Normative Ethics and Legal Theory*. Oxford: Clarendon Press.

Hooker, Brad. 2000a. *Ideal Code, Real World: A Rule-Consequentialist Theory of Morality*. Oxford: Oxford University Press.

———. 2000b. "Reflective Equilibrium and Rule Consequentialism." In *Morality, Rules, and Consequences*, edited by Brad Hooker, Elinor Mason, and Dale E. Miller, 222–38. Edinburgh: Edinburgh University Press.

Hristov, Alexander N., Kristen A. Johnson, and Ermias Kebread. 2014. "Livestock Methane Emissions in the United States." *Proceedings of the National Academy of Sciences* 111 (14): E1320.

Jackson, Frank. 1997. "Which Effects?" In *Reading Parfit*, edited by Jonathan Dancy, 42–53. Oxford: Blackwell.

Jamieson, Dale. 2007. "When Utilitarians Should Be Virtue Theorists," Utilitas 19 (2): 160–83.

Jamieson, Dale. 2014. *Reason in a Dark Time: Why the Struggle Against Climate Change Failed—and What it Means for Our Future*. Oxford: Oxford University Press.

Levy, Sanford S. 2000. "The Educational Equivalence of Act and Rule Utilitarianism." In *Morality, Rules, and Consequences*, edited by Brad Hooker, Elinor Mason, and Dale E. Miller, 27–39. Edinburgh: Edinburgh University Press.

Mason, Elinor. 2014. "Objectivism, Subjectivism, and Prospectivism." In *The Cambridge Companion to Utilitarianism*, edited by Ben Eggleston and Dale E. Miller, 177–98. Cambridge: Cambridge University Press.

Mill, John Stuart. 1969. *Utilitarianism*. In *The Collected Works of John Stuart Mill* Vol. X, edited by John M. Robson, 203–59. Toronto: University of Toronto Press.

Miller, Dale E. 2000. "Hooker's Use and Abuse of Reflective Equilibrium." In *Morality, Rules, and Consequences*, edited by Brad Hooker, Elinor Mason, and Dale E. Miller, 156–178. Edinburgh: Edinburgh University Press.

———. 2010. *J. S. Mill: Moral, Social and Political Thought*. Cambridge: Polity.

———. 2011. "Mill, Rule Utilitarianism, and the Incoherence Objection." In *John Stuart Mill and the Art of Life*, edited by Ben Eggleston, Dale E. Miller, and David Weinstein. 94–116. New York: Oxford University Press.

———. 2013. "Hooker on Rule Consequentialism and Virtue." *Utilitas* 25 (3): 421–32.

———. 2014. "Rule Utilitarianism." In *The Cambridge Companion to Utilitarianism*, edited by Ben Eggleston and Dale E. Miller, 146–65. Cambridge: Cambridge University Press.

Miller, Richard B. 2009. "Actual Rule Utilitarianism." *Journal of Philosophy* 106 (1): 5–28.

Miller, Scot M., Steven C. Wofsy, Anna M. Michalak, Eric A. Kort, Arlyn E. Andrews, Sebastien C. Biraud, Edward J. Dlugokencky, Janusz Eluszkiewicz, Marc L. Fischer, Greet Janssens-Maenhout, Ben R. Miller, John B. Miller, Stephen A. Montzka, Thomas Nehrkorn and Colm Sweeney. 2013. "Anthropogenic Emissions of Methane in the United States." *Proceedings of the National Academy of Sciences* 110 (50): 20018–22.

Parfit, Derek. 1984. *Reasons and Persons*. Oxford: Clarendon Press.

———. 2011. *On What Matters* Vol. I. Oxford: Oxford University Press.

Rawls, John. 1999. *A Theory of Justice*. Revised edition. Cambridge, MA: Belknap.

Ripple, William J., Pete Smith, Helmut Haberl, Stephen A. Montzka, Clive McAlpine and Douglas H. Boucher. 2014. "Ruminants, Climate Change and Climate Policy." *Nature Climate Change* 4: 2–5.

Ross, W. D. 2002. *A Theory of the Right and the Good*. New York: Oxford University Press.

Seibokaite, Aiste. 2015. "Climate Change as a 'Hard' Case of Collective Responsibility," in *Modern Dilemmas: Understanding Collective Action in the 21st Century*, edited by Dylan Kissane and Alexandru Volacu, 117–41. Stuttgart: *ibidem*-Verlag.

Scheffler, Samuel. 1994. *The Rejection of Consequentialism*. Revised edition. Oxford: Clarendon Press.

Smart, J. J. C. 1956. "Extreme and Restricted Utilitarianism." *The Philosophical Quarterly* 6 (25): 344–54.

———. 1973. "An Outline of a System of Utilitarian Ethics," in *Utilitarianism For and Against*. Cambridge: Cambridge University Press.

Smith, Holly M. 2010. "Measuring the Consequences of Rules." *Utilitas* 22 (2): 413–33.

U. S. Energy Information Administration. 2016. "Projected Growth in CO2 Emissions Driven by Countries Outside the OECD." https://www.eia.gov/todayinenergy/detail.php?id=26252, accessed 13 March 2018.

———. 2017. "Power Sector Carbon Dioxide Emissions Fall Below Transportation Sector Emissions." https://www.eia.gov/todayinenergy/detail.php?id=29612, accessed 31 March 2018.

van Kooten, Cornelis G. 2017. "Forest Carbon Offsets and Carbon Emissions Trading: Problems of Contracting." *Forest Policy and Economics* 75: 83–8.

World Bank. no date. "CO2 Emissions (Metric Tons Per Capita)." https://data.worldbank.org/indicator/EN.ATM.CO2E.PC, accessed 13 March 2018.

5

KANT AND CLIMATE CHANGE

A Territorial Rights Approach

Alice Pinheiro Walla

Kant and Climate Ethics

Anthropogenic climate change poses a considerable threat to human life on planet Earth. Extreme weather, water stress, crop failure and the spread of diseases are among the effects of climate change that are already being felt around the globe. However, climate change does not pose a threat to *all* life: nature can certainly adapt to new conditions. The question is if *we* can adapt. Since global warming is already a reality and can no longer be averted, addressing climate change involves two types of measures: *mitigation* (attempts to curb and regulate greenhouse gas emissions and restrain individual consumer lifestyles) and *adaptation* to new living conditions (financially supporting farmers affected by droughts, relocating residents threatened by rising sea levels, addressing new epidemics and climate migration). Sadly, experience shows that the more vulnerable the population affected by the adverse effects of climate change, and the greater the danger to their subsistence and physical integrity, the less the political will to support them.

Kant's ethical theory provides normative guidance concerning our duties to others and to ourselves. It is no surprise that climate ethics has also turned to Kant and to Kantian inspired theories in search of normative principles for dealing with the challenges of climate change and spelling out what we owe to each other in this regard. While I do not dispute that there are important individual ethical duties in regard to climate change, the problem is that the focus on individual virtue and individual agency seems inadequate in the face of the urgency to mitigate the impact of climate change. Although I propose in this article a Kantian approach to climate change, I will start by explaining the limitations of applying Kant's *ethical* theory to the problem of climate change. Climate change is a global problem that calls for urgent collective action not only at the domestic level but also at the international level. Therefore, applying a theory of individual duties to a collective, global problem is problematic. Either individuals are expected to *compensate* for the lack of juridical-political institutions with individual virtuous conduct[1] or we are applying a theory of individual ethics to political institutions. In fact, the application of Kant's ethics to political problems has been the rule in Kantian political philosophy, since Anglophone Kantians were barely acquainted with Kant's legal-political thought.[2]

After a brief sketch of the main tenets of Kant's ethical theory, I will analyse an interpretation of the categorical imperative as a principle of sustainability that could be applied to the issue of climate change. However, I refute this interpretation due to its problematic understanding of the universalisation procedure expressed by the categorical imperative. I then argue that using Kant's legal theory is more promising for addressing climate change due to its ability to justify externally enforceable duties and its focus on political-juridical institutions. Next, I explain how climate change can constitute a hindrance to external freedom from the perspective of Kant's legal philosophy. Because all persons have an equal original right to be somewhere on the Earth, territorial rights and border control are only permissible if those excluded have somewhere else to be. Climate change, however, poses a threat to the livelihoods and existence of the most vulnerable in the planet, who are then prevented from relocating due to the world's configuration into territorial nation states. This gives rise to specific duties of territorial states to address climate change at a global level. Finally, I conclude by making the case for securing the territorial rights of indigenous peoples for the purpose of environmental preservation and providing an alternative model for future generations.

The Categorical Imperative as a Principle of Sustainability

Kant's moral theory seems attractive for accounting for our duties to others in regard to anthropogenic climate change. Kant's moral theory is a duty based morality, meaning that obligations (rather than rights) are the fundamental ethical category.[3] Further, it is principle based: according to Kant, the categorical imperative is the principle underlying all our (sincere) moral judgements, whether we are aware of this principle or not.[4]

Kant's categorical imperative principle arises from the idea that morality can only be binding if its fundamental principle can be conceived as arising from the agent's own will. If the will is free, so Kant's argument goes, then it must not be determined by external forces; but if the will is also a *causality in the world*, it must also operate in accordance with laws. From this Kant concludes that a free will must be the *source* of its own fundamental law. Kant then shows why the categorical imperative, a formal principle imposing constraints on our subjective principles of actions, which he calls *maxims*, is the only principle compatible with a free will. Action on the basis of inclinations presupposes principles that are normative *only* because they serve pre-existing inclinations of the agent. Kant calls these conditional rational principles "hypothetical imperatives." In contrast, the moral law must be normative regardless of what we happen to want and desire at a given moment. This view of morality is also in line with our non-philosophical, everyday understanding of *duty*; that is, the view that certain actions are morally necessary regardless of our inclinations (GMS IV: 397–401). Only conduct that is properly motivated by respect for the moral law is strictly speaking *moral* (and consequently *free*). A moral theory that does not recognise the autonomy of the will as the source of morality would necessarily conceive moral requirements as being *external* to the agent's will. But this, Kant argues, would be incompatible with the idea that our will is free and that morality is the very expression of the freedom of the will. Therefore, morality must be conceived as arising from the autonomy of the will and its law must be purely *formal*: a requirement that our maxims (our subjective principles of volition) must have certain formal properties. Although maxims themselves are subjective principles expressing what we are setting ourselves to do and involving an implicit conception of the context and particular circumstances in which we are situated as agents, morality requires that maxims be nevertheless *fit* to become universal laws. That is, in order to be morally permissible, our maxims must be such as to be conceivable and willed as universal principles. "Act only according to that maxim through which you can at the same time will that it become a universal law" (GMS IV: 421).

Kant also offers other formulations of the categorical imperative, which are supposed to be equivalent to the first "universal law of nature" formulation (GMS IV: 421). The second formulation, prohibiting the instrumentalisation of rational agents (GMS IV: 427–428), is particularly famous for expressing the Kantian idea of respect for persons, often associated with the concept of dignity.[5]

How can the categorical imperative be applied to the problem of climate change? Martin Schönfeld proposed a naturalistic interpretation of the categorical imperative, according to which moral values mirror natural facts about the world and human nature. He argues that the categorical imperative expresses a requirement of universality that is similar to the idea of *sustainability*; just as universal lying would undermine communication, ultimately rendering particular instances of lying impossible, environmentally unsustainable maxims necessarily undermine themselves. This is why the categorical imperative can be regarded as the "naturalistic blueprint of sustainable activities" (Schönfeld, 2008, 50). According to Schönfeld, the requirement to universalise one's maxims has the same inner logic of environmental systems: an action is rationally defensible if others can imitate it without creating a situation which would preclude future action: "as a rule, an action is wrong if its replication entails the end of replication" (Ibid., 56). It follows that morally permissible action is universalisable by virtue of being self-sustaining. A maxim of unrestrained consumption of the Earth's resources would undermine itself and therefore be impermissible.

While this naturalistic interpretation seems at first to be only compatible with the first formulation of the categorical imperative (the so called "formula of law of nature"), Schönfeld also applies it to the two other formulations. The second formula, commanding respect for humanity as an end in itself, is interpreted as requiring that persons be treated in accordance with their natural constitution, that is, their autonomous nature. Schönfeld argues that people naturally grow up to be independent and autonomous, and that to treat persons as ends in themselves is none other than treating them "as they are," that is, as "potentials in time." Moral values thus reflect natural facts about persons, not only facts and regularities in the world.

If we conceive each self-legislating agent as determining her own volition in accordance with a universal law, it is possible to think of a resulting systematic connection among all rational beings. Schönfeld sees in the third formulation of the categorical imperative (autonomy in a kingdom of ends) a future-oriented aspect of Kant's universalisation requirement:

> By a kingdom [...] I understand the systematic union of several rational beings through common laws. Now, since laws determine ends according to their universal validity, it is possible — if one abstracts from the personal differences among rational beings, and likewise from all content of their private ends — to conceive a whole of all ends (of rational beings as ends in themselves, as well as the ends of its own that each of them may set for itself) in systematic connection, i.e., a kingdom of ends, which is possible according to the above principles.
>
> *GMS IV: 433*

A kingdom of ends would be *possible* if our maxims satisfy the formal requirements of the categorical imperative. Schönfeld concludes that the kingdom of ends formulation must also include future generations and requires us not to discount against the future (Schönfeld, 2008, 58).

Despite the parallels between the idea of sustainability and the categorical imperative's universalisation requirement, Schönfeld's interpretation is problematic. First, one can question the naturalistic interpretation of the categorical imperative. Second, Kant's moral theory is about individual willing; that is, the quality of the agent's maxims. One may question whether the focus on individual morality is helpful when it comes to influencing the climate system. This latter objection is not restricted to Schönfeld's account, but would apply to any application of Kant's *ethical* theory to climate change.[6]

How should we understand the universalisation requirement expressed by the categorical imperative? There is a tendency to interpret the categorical imperative as asking a question of the form "I want to φ, but what would happen if everybody φ-ed?"[7] This simplistic interpretation has given rise to many misunderstandings of the universalisation procedure such as the objection that wanting to become a doctor or going through a door before someone else would fail the categorical imperative test although intuitively these are not necessarily immoral maxims.

I will argue that Kant understood universalisation as a requirement of *principle consistency*. The rationale of the requirement is not to show that I would not be able to act on my maxim if everybody did the same. Why should the thought of everybody acting like me deter me, since this is only an imaginary universalisation scenario? Instead, Kant's universalisation test reveals that the agent is making an *exception* for herself from a principle she must necessarily want others to adopt; in fact, general compliance with the principle is what makes her rule breaking possible. The procedure thus exposes the agent's *double standards*: all others should abide by the principle while I make an exception for myself. Therefore, Kant's point is not that I will not be able to get what I want if my maxim cannot be universalised. As immoral agents know well, immoral conduct such as deceiving, cheating, manipulating and so on are often the easiest and most efficient means to quick results. Kant's point is that making an exception for yourself from a principle you must want all others to respect is *immoral*. Why? Because *it violates reciprocity*. It shows that I want others to be bound by a principle I am not willing to bind myself to. I am making an exception for myself while everyone else should play by the rules. And *that* is immoral.

Kant identifies two ways in which a maxim can fail to be universalised. The first and more straightforward way is illustrated by the lying promise. The contradiction in that case is known as "contradiction in conception" because I cannot even *think* of my action as being possible in a world in which my maxim were universalised. If lying promises were the rule, I would never be able to help myself with such a strategy, since no one would believe me in the first place. But as Kant concedes, there are maxims that can very well be conceived as universal laws without undermining themselves and are nevertheless immoral. These maxims would be "sustainable" but not compatible with a substantive ideal of rational agency. Kant's example of a sustainable but immoral maxim is indifference to the plight of others. Imagine someone who adopts a maxim not to violate any rights, but not to help others either (with the willingness not to be helped in return). The agent is not free riding on the good will of others or making exceptions for herself; she doesn't help anyone but claims no one's help. This may sound like a fair deal as a form of reciprocity in a *negative sense;* I don't do anything for you and you don't need to do anything for me. The universalisation of my maxim of indifference would not undermine the possibility of ignoring the needs of others. My maxim would not undermine its own realisation when universalised. Why is it then immoral?

A maxim like this is not compatible with *rational agency as such*. It is not that it would be silly or dangerous to deprive myself of the help of others when I am in need (this would be a maxim of *prudence* rather than morality); the idea is that no rational agent could possibly *want* to voluntarily deny herself the possibility of being helped by others as a matter of principle, in return for not having to care about anyone else. This second type of contradiction, known as "contradiction in the will," is more controversial than the first "contradiction in conception" because it does not rely on a requirement of reciprocity, but on a substantive conception of rational agency. Rational agency is about setting oneself ends and pursuing them; therefore, it cannot be rational to *choose* to deprive oneself of the possibility of achieving one's ends. Adopting a maxim of beneficence towards others is a duty because the opposite maxim is impermissible (GMS IV: 423).

Schönfeld interprets the categorical imperative as a requirement of *prudential* consistency: if I want my action to be possible in the future, I must adopt only environmentally sustainable maxims. If I want to be able to use this resource in the future, I must not deplete it. This interpretation can only give rise to what Kant called "hypothetical imperatives;" that is, imperatives of the form "if you want x, then take the necessary means y." Hypothetical imperatives presuppose that we are already committed to the end in question (because we happen to desire or want it) and command us merely to take the necessary means to that end's realization. This might work in the case of environmental policies if we are *already* motivated and committed to pursue environmental goals. But it cannot command us to adopt the end *categorically*, if we do not care about it. And this is precisely the advantage of the categorical imperative, as Kant understands it: it is a principle that commands unconditionally, regardless of one's desires and interests because it imposes obligations that are unconditionally binding.

Another problem with Schönfeld's interpretation is that such a requirement only yields the intended results if every individual is *equally committed* to the adoption of sustainable maxims. If I am the only person adopting sustainable maxims while everyone else is polluting and exhausting natural resources, I can hardly make a difference. In contrast, the bindingness of Kantian duties does not depend on the compliance of others; even if no one else is acting morally, it is still the case that the moral law is binding for me. While ineffectiveness may be no argument against moral obligation, it is questionable when it comes to tackling climate change. After all, we do want to change the world in this regard. I conclude that sustainability is a metaphor for a deeper moral intuition in Kant's moral theory, and not a moral requirement *per se*.

One could argue that the greatest limitation of Kant's ethical theory in regard to climate ethics is that it is *anthropocentric*. It is based on respect for rational nature in our person and in the person of others and therefore excludes the possibility of *direct* obligations towards the environment. However, Kant's ethical theory does not leave us with a deontological "vacuum" in regard to the environment. He accounts for duties towards non-human animals and nature in general as *indirect duties* to oneself (MS VI: 442). Cruelty to animals and wanton destruction of nature not only have a bad influence on one's moral character but they also reveal an immoral disposition. Insensitivity to beauty in nature and to the destruction of the environment reveals something very deep about oneself. A person who is cruel to animals is hardened against suffering in general. Although Kant's argument here seems to be that she poses a danger to other human beings (and this is what's bad about cruelty to animals), I see Kant as suggesting a point about moral psychology: our dispositions must be *consistent* in regard to all sentient beings. Our emotional responses to recognition of suffering play an important role for moral judgement and moral character. Drawing a sharp line of concern between rational and non-rational sentient beings would lead to an inconsistency in our emotional responses. Although it is our duty in regard to non-human animals to refrain from cruelty and even to express *gratitude* to old domestic animals for their services (MS VI: 442), the *justification* of the duty lies ultimately in the agent herself, in her rational nature. We must thus distinguish between our attitude towards the beneficiaries of the duty, including our responsiveness to suffering in general, and the justification of the duty in rational nature. The way Kant justifies obligation as based in reason does not preclude concern for non-rational sentient beings; on the contrary, it suggests that our relation to non-human beings is the very *locus* for the development of those emotional responses, since this can be done without the danger of paternalism and unintentionally humiliating the other.[8]

Nevertheless, I will argue that the problem with applying Kant's moral theory to climate change is the *nature* of the obligations which arise within the context of climate change. Climate change affects livelihoods and people, who in turn become dependent on the assistance of others in order to readapt to the new circumstances. This is not to say that individuals do not have ethical duties to do their share in mitigating climate change and to assist those in need. Therefore, my claim is not that applying Kant's moral theory to climate change leads us to wrong conclusions. Instead, the problem is that ethical duties are not externally enforceable and therefore are fully dependent on the good will of individual agents.

Since climate change impacts people across the globe and calls for urgent collective action, it is questionable whether focusing solely on individual virtue is an adequate approach to the challenge in question. Since not everyone can be expected to fulfil their obligations, we also need to reflect about the limitations of a purely ethical, individual approach to a collective, global problem and what individuals and collective agents can be rightfully *coerced to do*. This reflection, I believe, is provided by Kant himself, when he introduced a distinction between the domain of individual virtue and the domain of coercible juridical obligations in his late work *The Metaphysics of Morals* (1797). This distinction was not yet present in Kant's earlier, foundational ethical works, the *Groundwork* and *The Critique of Practical Reason*.[9] As Kant explains in the introduction to *The Metaphysics of Morals*, the difference between the domain of Ethics or Virtue (*Tugend*) and the domain of Right (*Ius, Recht*) lies in their respective kind of *lawgiving* (*Gesetzgebung*). Ethics and Right connect different incentives to the law:

> That lawgiving which makes an action a duty and also makes this duty the incentive is called *ethical*. But that lawgiving which does not include the incentive of duty in the law and so admits an incentive other than the idea of duty itself is *juridical*.
>
> *MS VI: 218*

The incentive in ethics must be *respect* for the moral law. The constraint in this case must be self-imposed and thus *internal*: it does not make sense to coerce someone to be moral, since moral motivation is something that only the agent herself can do. One can only be coerced to external actions, not to *willing* a principle. In contrast, the domain of Right does not require ethical motivation: mere external compliance with laws is sufficient. Since Right does not require ethical motivation, it is possible to coerce and provide external incentives for compliance with juridical laws (for instance, sanctions).[10]

Climate change is a global phenomenon involving the actions of many individuals and affecting many others. Addressing climate change will thus require external regulation and coordination at global level. This calls for more than individual will: collective, that is, *political* action. Climate change thus raises the question of what kinds of actions we can *impose* on others externally, rather than merely appealing to their conscience. Actions that can be externally coerced are the object of Kant's legal theory.[11]

Kant's Legal Theory and Climate Change

Kant's legal theory is concerned with the regulation of external interactions between persons, so that individual spheres of external freedom and subjective rights are mutually compatible and protected. While ethics has to do with internal freedom, that is, with the *quality* of the agent's willing, external freedom concerns the possibility of physical interaction between embodied persons in space. A person's choices have an external impact on the choices of others. The task of Right (*Recht*)[12] is to make these choices compatible with each other. As Kant puts it, Right "is the sum of the conditions under which the choice (*Willkür*) of one can be united with the choice of another in accordance with a universal law of freedom" (MS 6: 230).

One may think that, from a Kantian perspective, morality fully answers the question of which actions we can impose on others. "Treat others as ends in themselves and never as mere means" would be all the guidance we need in our relations with others. However, Kant's legal theory reveals a more complex picture. Not everything one does to others *externally* is in accordance with Right, even if one's underlying intentions are morally good. To use an explanation offered by Hart, one must imagine each person as having an equal sphere of freedom (Hart, 1995). Our task is to regulate the relations among those equal spheres of freedom, thereby integrating them in a coordinated *system*. Moral motivation alone cannot secure this coordination, since the task at hand requires more than the quality of my volition. Having a plurality of well-meaning drivers does not secure a well-coordinated traffic system. It is not a matter of individual character or good will, but of external regulation.

Since we share the planet with each other, interaction is unavoidable; sooner or later the impact of our actions will be felt by others. Climate change perfectly illustrates the fundamental problem of Right: actions on one part of the planet are affecting persons in other locations. But this impact must be understood in terms of *external freedom*, or precisely as a 'violation' of external freedom. Since one can easily trespass on the sphere of external freedom of another person, our interactions must be regulated to preserve each person's equal sphere of freedom. As Hart puts it, we must maintain an "equal distribution of restrictions and so of freedom" among persons (Ibid., 191). For this, Kant proposes a formal principle requiring that our capacity to impose an obligation on each other be *symmetric*. Normatively speaking, I am unable to bind you in a way you could not bind me in return. Otherwise we would not be equals in juridical status; one of us would be free and the other a slave. A slave is someone who has no say in regard to her external freedom; that is, she cannot *bind* other persons to respect her external freedom.

However, one may also choose to restrict one's own sphere of freedom, such as in contracts. You may acquire a claim to a certain action or performance of mine that I do not have on you. I would be binding myself to you in a way you are not bound in return. The reason this is still compatible with *Recht* is that I *chose* to enter a contract with you. It would be problematic if your interference on my sphere of freedom has nothing to do with my consent, which is required since I am the one with the *authority* to decide over my own sphere of freedom. But there are things one may impose on me, whether I want it or not. The central idea of Kant's legal philosophy is that reason can provide the principles that will ensure that nothing is externally imposed on others *arbitrarily*, that is, without the *authority* to do so. This fundamental principle of reason is the principle of right. "Any action is right if it can coexist with everyone's freedom in accordance with a universal law, or if on its maxim the freedom of choice of each can coexist with everyone's freedom in accordance with a universal law" (MS 6: 231).

It is important to note that Kant is talking about *Willkür* (usually translated as "choice") and not about *Wille*, as in his moral theory. This is a significant point, since Kant associates *Wille* with the noumenal self and transcendental freedom and *Willkür* with the phenomenal self, that is, ourselves as appearances in time and space. Since we are regarding ourselves as phenomena, as embodied beings in space, the internal quality of one's maxim is *secondary*; what is important is that one's action can coexist externally with the equal freedom of choice of another; that is, *we do not violate another's authority over her own domain of external freedom*. Imagine I took away your coat to give to someone who needs it more. After all, you have many coats and don't even seem to care about that one. However, my good intention, the greater need of another person, and your abundance of resources do not allow me to take away your coat, which is your property; that is, something belonging to your sphere of external freedom.

This may sound terribly wrong. How can one defend the right to private property while allowing the more fundamental needs of others to be neglected, especially when the property in question seems such a trifle to its wealthy owner?

It seems morally intuitive that private property should be conditional on moral considerations permitting or forbidding the control of a certain resource or object depending on the circumstances. But this is not the point. The idea is that, formally speaking, I am imposing something on you I have no *authority* to do, even if I have the best moral reasons on my side. I am transgressing into your sphere of external freedom. This does not mean that private property is absolute. For instance, Kant argues that the state has the *right* to tax the wealthy to maintain those unable to provide for themselves. As an individual I have no authority to impose this on you, but the state, as a public institution, does have this authority:

> The wealthy have acquired an obligation to the commonwealth since they owe their existence to an act of submitting to its protection and care, which they need in order to live; on this obligation the state now bases its right to contribute what is theirs to maintaining their fellow citizens.
>
> *MS VI: 326*

The state has the *authority* to redistribute (taxing the wealthy) while individuals living together in a civil condition cannot impose this on each other. The question is thus not whether property rights are absolute but *what* can be externally imposed and *who* has the authority to impose it. Kant observes that if an action violates external freedom, hindering that action would be compatible with right (MS VI: 231–232). The possibility of coercion is thus implicit in the very concept of a right. An action that cancels a violation of external freedom would actually be right ("a hindrance of a hindrance to freedom"). But this does not mean that whoever *has a right* also has the authority to enforce that right. One also needs the *authority* to coerce. One may agree with Locke that anyone who "is in the right" has such an authority. Kant denies that individuals have the authority to coerce each other externally even though their rights are violated. This follows from their radical equality. Making justice with one's own hands is indeed inevitable in the state of nature. The state of nature is a situation in which individuals must follow their private judgements about rights (and their own conscience, as in ethics). But even if we assumed that individuals will act in their best moral knowledge and conscientiously, they are acting *unilaterally*. As *private persons* individuals lack the authority to coerce. For this we need political representation and public institutions. We thus have a duty to leave the state of nature and are exceptionally authorised to coerce others for that purpose (MS VI: 307; Pinheiro Walla, 2014).

Climate change is caused by the *externalities* of our current lifestyles. Whether these actions were *intentionally* harmful is irrelevant to *Right*. What is significant is whether they impact on the external freedom of other persons in a way that is incompatible with their equal external freedom. I have argued that Kant's legal theory is more promising than his ethical theory when it comes to climate change because it is concerned with duties that can be externally coerced and does not require virtuous motivation from agents. However, one may wonder if Kant's legal theory is truly helpful when it comes to climate change policy and regulations. Since external freedom is at the core of Kant's legal theory, one would have to show that environmentally harmful practices are incompatible with external freedom. However, changes to the environment *per se* do not seem to count as violations of external freedom (Ataner, 2012, 142). Arthur Ripstein notes that if a person acquires one object ahead of another, say, the last quart of milk available at the supermarket, she is not violating the right of the person who came later and was left empty-handed. This is because the person who came after her had no claim to that last quart of milk. There was no interference with her capacity to exercise external freedom (Ripstein, 2009, 16). Are the climate conditions we are now generating to be understood in the same way as Ripstein's last quart of milk scenario, namely, as a change in the environment that does not amount to an interference with one's capacity to exercise external freedom?

First we must ask ourselves what counts as an interference with one's capacity to exercise external freedom, in other words, what amounts to a violation of right from the perspective of Kant's legal philosophy. External freedom can only be restricted by some form of external hindrance. But not all forms of external hindrances are unlawful; in fact, coercion may be justified if it protects external freedom from arbitrary interference (MS VI: 231). Further, Kant's legal theory is *relational*. A state of affairs that constrains the scope of my choices is *per se* not necessarily an unjustified interference of my external freedom. A violation of external freedom is the result of another's exercise of external freedom overstepping the boundaries of reciprocal external relations. In other words: someone must be violating another's equal entitlement to freedom from interference in such a way that she is imposing a constraint on another that she is herself not subjected to. She is therefore violating the reciprocity requirement that is fundamental for legal relations; that is, negating another person's equal juridical standing. For my argument, I need to show that climate change amounts to an unlawful restriction of external freedom in the relevant sense, that is, climate change does not merely impact on people's freedom but is incompatible with the equal juridical standing of all persons as members of a global community.

Elsewhere I have argued that Kant based both the territorial rights of states as well as the right to be admitted in a foreign territory in case of necessity on the same principle, namely, original community of land (*communio fundi originaria*) (Pinheiro Walla, 2016). According to Kant, persons have an original right to be wherever nature or chance (*apart from their will*) has placed them (MS VI: 262, my emphasis). Originally, no one had more right than another to be on a place on the Earth (ZeF VIII: 358). Although one can acquire a right to control territory due to prolonged possession of land, this right is not absolute. While host countries maintain the right to refuse *voluntary* interactions under the condition that they do not treat peaceful visitors with hostility, they have no right to refuse those whose interaction is *involuntary*. If all persons have an equal original right to exist on the planet, and no one has the authority to bind others in a way others could not bind her in return, one cannot be obligated to respect the occupation of space of others when this entails her being deprived of a place in which she can exist.

Depletion and destruction of specific natural resources can amount to a direct violation of external freedom. For instance, when irresponsible exploitation of natural resources by a mining company leads to the poisoning of a local community's water sources and soil. The violation in this case consists not only of a negligent destruction of the community's means of survival within the territory they occupy, but is also a violation of their physical integrity, given the threat the pollution poses to their health, wellbeing and livelihoods. However, arguing that climate change *in general* amounts to a violation of external freedom is more complex, since identifying perpetrators and quantifying their specific impact on those adversely affected on other parts of the globe is difficult, if not impossible. It is certain, however, that the livelihoods of communities from poorer areas of the planet are being threatened by climate change. Where adaptation costs cannot be met locally, migration may be the only viable solution. This is where territory becomes relevant to the philosophical debate on climate change.

Borders impose constraints on the mobility of individuals and limit their access to opportunities. In some cases, even the basic ability of affected persons to provide for themselves may be compromised if they are prevented from trying their luck elsewhere. The "negative" answer to the threat of climate change would be thus unrestricted global freedom of movement for climate change refugees. This would enable a geographical redistribution of the Earth's population away from areas rendered inhabitable, inhospitable or with no adaptation resources. The problem with the negative solution is that today's global configuration in territorial states does not allow such global mobility to the world's poor, even less global mobility *en masse*. A "positive" alternative, provided by wealthier states bilaterally in order to avoid mass migration is financial aid for adaptation within the affected territory.[13]

Therefore, while climate change is not a *direct* violation of external freedom, closing off adaptation possibilities to climate change refugees is. This is because of the asymmetry in the way restrictions to external freedom are distributed. Those most severely affected by climate change are also most restricted in their possibility to adapt. Occupying territory is permissible under the assumption that other people have somewhere else to be and will not be under life threat if denied entrance. The natural conclusion would be that climate change would require states to open their borders to those affected by climate change, allowing a geographic redistribution of the Earth's population for the purpose of adaptation to the new global conditions. But states are *unwilling* to admit foreigners into their territory. It follows that territorial statehood gives rise to special obligations towards those affected by climate change, especially climate change refugees and climate change migrants *by virtue of their territorial rights*. The global community of states has thus a special duty to provide a satisfactory alternative arrangement to unrestricted global mobility. This would involve not only a duty to mitigate climate change, but also a duty to provide assistance to those affected by climate change to adapt to the new global conditions.

Indigenous Territorial Rights: Preservation and Imagination

I argued that states have a juridical duty to implement global regulations to address climate change by virtue of their control of territory and restriction of the global mobility required for climate change adaptation. But is this helpful, given the urgency of the problem and the slow pace of international action? Despite the warnings of scientists and activists, states are too slow and reluctant to take action. In this final section, I will suggest one possible way to address climate change which seems more feasible in the face of international political inaction, because it is *conservative*.

Much of the climate change discussion in ethics is focused on the moral justification for curbing carbon emissions and the fair distribution of the costs of climate change mitigation and adaptation (Broome 2012; Caney 2009). This approach takes for granted our current lifestyles and conception of what land is and how to relate to it. The question is how to keep climate change under control and distribute burdens and costs between differently situated global actors. In the meantime, indigenous groups in different parts of the globe are struggling to protect their ancestral lands against allegedly more urgent energy and economic needs. Indigenous peoples' territories cover around 24% of all land worldwide and hosts 80% of the planet's biodiversity (UNPFII factsheet, no date). And yet, because their lands are resource rich and they often suffer discrimination within their respective legal systems, indigenous peoples are being systematically dispossessed and murdered. Indigenous women are particularly at risk (Permanent Forum on Indigenous Issues, 2019).

Securing indigenous land rights would be a way to preserve the remaining biodiversity of the planet, provided indigenous groups remain committed to land stewardship and sustainability. Therefore, environmental regulations should also apply to their use of the land. Since this is a conservative strategy, it is arguably less demanding than setting up still inexistent international institutions with adequate coercive powers or even getting global actors to agree to take action, although it is not meant to *replace* these measures either. Indigenous peoples can also provide an invaluable contribution to the climate change debates and policies. As Krushil Watene and Many Yap argued, focusing on Māori and Aboriginal peoples, indigenous peoples can significantly contribute to sustainable development goals with their traditional knowledge in the management of land, water and natural resources, and even with their specific cultural values, but are rarely included in such discussions. They also stress the need of indigenous self-determination for the possibility of such contributions (Watene and Yap, 2015, 51–55).

The existence of indigenous peoples and groups who relate to land in non-exploitative, respectful and sustainable ways is for all nations on the Earth and for future generations both an important alternative model and an ideal for long-term change. This is because being aware of the possibility of relating to land otherwise, that is, with emotional attachment, local knowledge and respect (where these relations to land still exist) enables us to question and in the long term revise the way we relate to the planet, which otherwise may be taken for granted. This is in fact a Kantian idea. Kant argues that in certain cases radical political reform may be too hasty (ZeF VIII: 347). This may be because the people's mentality is not ready for change or for other pragmatic reasons. In this case, we should not attempt to implement change at any costs, since this could undermine the end we are trying to promote. It is therefore permissible to postpone reforms provided the end *is not made impossible* in the future.[14] In the same vein, although it is unrealistic to expect a radical change in our present relation to land, the existence of alternative models can help us reflect deeper about own ways and *keeps open* the possibility of change, especially for future generations.

In this chapter, I explained why applying Kant's ethics to climate change has limitations. I proposed applying Kant's legal-political philosophy and considering the territorial aspect of the problem of climate change for understanding our global duties in regard to climate change. I suggested that respecting the territorial rights of indigenous peoples would be a possible way to protect areas which are still untouched by environmental destruction, and that it would also provide an alternative ideal for possible long-term change.

Since the institutions able to enforce climate regulations and effectively protect indigenous territorial rights are still nonexistent and we still depend on the willingness of states to put them in place, we must rely on individuals to put pressure on their national governments and call for their implementation. In other words, we are still addressing rights violations as matters of individual virtue. But this does not mean that our duties to each other in the face of global warming are only duties of virtue. I argued that the territorial character of nation states and their desire to protect borders and prevent global migration as a means to climate change adaptation give rise to stringent duties of right to mitigate climate change and provide adequate support to those worst affected by it. Addressing the global issue of climate change is thus required by the equal juridical status of all persons in a world of hard borders and territorial states.[15]

Notes

1 This would also explain why trying to tackle climate change individually (by radically changing one's lifestyle and perhaps engaging in activism) could become extremely demanding for the individual in question, depending on the external circumstances. The source of moral demandingness arises in this case from the fact that individuals would be compensating for the lack of political institutions with their personal conduct. On the other hand, one could argue that individuals have a duty to implement and promote juridical-political institutions.

2 This "applied ethics" approach was famously adopted by John Rawls. Given the growing number of scholarly works on Kant's legal-political thought in the last twenty years, this area of Kant's practical philosophy has been receiving more attention in political philosophy (see for instance Byrd and Hruschka, 2010; Ellis, 2005; Ripstein, 2009; Flikschuh, 2000).

3 Kant's works are cited in accordance with the volume and page numbers of the standard edition of Kant's works (*Akademieausgabe*). I use Jens Timmermann's facing-page edition of the *Groundwork* and his revised version of Mary Gregor's (1996) English translation. All other translations are from *The Cambridge Edition of Kant's Works*, published by Cambridge University Press. I use the following abbreviations:

GMS Groundwork of the Metaphysics of Morals
MS The Metaphysics of Morals
ZeF Towards Perpetual Peace

4 It is not Kant's intention to provide a new principle for morality. The categorical imperative is supposed to be the formulation implicit in everyday moral reasoning, but in obscure form (GMS IV: 403–404). See also Timmermann (2007, xii).

5 It is noteworthy that the concept of dignity actually appears for the first time in the *Groundwork* in connection with the third formula, "Autonomy in a Kingdom of Ends," and not with the "Humanity as an end it itself" formula. See GMS IV: 433–34.

6 As I will explain later, Kant distinguished between the ethical and the juridical domain in his later works. While his ethics has to do with individual duties of virtue, the juridical domain covers duties that are externally enforceable. I will argue that Kant's legal-political philosophy may be a more promising aspect of his theory for addressing climate change than an application of his ethical theory.

7 This oversimplified understanding of the rationale of the categorical imperative procedure is the basis of the famous Sidney Morgenbesser Kant anecdote:

> [Morgenbesser] was smoking in the subway. A transit cop came up to Professor Morgenbesser and demanded that he put out his pipe. "What if everyone smoked?" the cop said, reprovingly. "Who are you—Kant?" the irritated professor asked, whereupon the policeman, misunderstanding "Kant" as something else, hauled Sidney Morgenbesser off to the precinct house.
>
> *Denby, 1996, 250*

8 When discharging our duties towards other persons, especially meritorious duties or duties of virtue, it is particularly important to avoid humiliating the beneficiary, who is put in a situation of indebtedness to the benefactor (MS VI: 453, §31). Also, when helping others, it is important to take into account their permissible ends and not attempt to make them happy against their wishes and conception of happiness (MS VI: 388). This worry does not arise in regard to animals, although we must bear in mind their species-specific wellbeing.

9 For an account of the development of Kant's practical philosophy, see Wood (2002).

10 Kant notes that although the juridical domain does not require it, it is still possible to comply with juridical duties with ethical motivation:

> It is no duty of virtue to keep one's promises but a duty of right, to which one can be coerced. But it is still a virtuous action (a proof of virtue) to do it even

> where no coercion may be *applied*. The doctrine of right and the doctrine of virtue are therefore distinguished not so much by their different duties as by the difference in their lawgiving which connects one incentive or the other with the law.
>
> *MS VI: 220*

11 One could argue that it is the ethical duty of individuals to create the institutions which will uphold and enforce our duties to others (for instance, the duty to support those who cannot provide for themselves). Individuals would thus transfer at least some of their ethical duties to more efficient institutions, which would also help to take the burden off each individual's shoulders. But there is a question we are forgetting to ask: what can institutions *impose* on people? Certainly not the enforcement of ethical duties *qua ethical duties*. This would amount to paternalism from a Kantian perspective. Institutions can only externally impose what is compatible with the external freedom of individuals. This is the view I develop in the next section.

12 I will be using the awkward term "Right" as a translation of the German "Recht" for lack of a better term. *Recht* is closer in meaning to the idea of rule of law than to justice or law.

13 This has been US policy in Central America. For instance, financial aid has been provided to coffee farmers in Guatemala in the hope of curbing immigration to the US (Semple, 2019).

14 This also applies to Kant's theory of peace. Even though peace may not be achievable in our lifetime, it is important to preserve the conditions under which a future peace would not be precluded. Kant provides principles for the achievement of peace in the form of articles in *Towards Perpetual Peace*.

15 I would like to thank Ben Eggleston and Dale Miller for their invaluable feedback on earlier versions of this article and audiences at Tel Aviv University, Heinrich-Heine University of Düsseldorf, and at the University of Oslo for very helpful discussions.

Bibliography

Allen Wood, 2002, "The Final Form of Kant's Practical Philosophy" in Mark Timmons (ed.), *Kant's Metaphysics of Morals*, Oxford: Oxford University Press, 1–22.

Ataner, Attila. 2012. *Kant on Freedom, Property Rights and Environmental Protection*. http://hdl.handle.net/11375/12678

Broome, John. 2012. *Climate Matters: Ethics in a Warming World*, New York: W.W. Norton.

Byrd, B. Sharon and Hruschka, Joachim. 2010. *Kant's Doctrine of Right*. Cambridge: Cambridge University Press.

Caney, Simon, 2009."Justice and the distribution of greenhouse gas emissions." *Journal of Global Ethics*, 5:2.

Denby, David. 1996. *Great books: My adventures with Homer, Rousseau, Woolf, and other Indestructible Writers of the Western World*, New York: Simon & Schuster.

Ellis, Elisabeth. 2005. *Kant's Politics. Provisional Theory for an Uncertain World*. New Haven and London: Yale University Press.

Flikschuh, Katrin. 2000. *Kant and Modern Political Philosophy*. Cambridge, UK: Cambridge University Press.

Gregor, Mary (trans. and ed.). 1996. *The Cambridge Edition of the Works of Immanuel Kant. Practical Philosophy*. Cambridge: Cambridge University Press.

Hart, H. L. A. 1995. "Are There Any Natural Rights?" *The Philosophical Review* 64, 175–191.
Krushil Watene and Mandy Yap, 2015. "Culture and Sustainable development: indigenous contributions." *Journal of Global Ethics* 11(1), 51–55.
Pinheiro Walla, Alice. 2016. "Common Possession of the Earth and Cosmopolitan Right." *Kant-Studien* 107(1), 160–178.
Pinheiro Walla, Alice. 2014. "Human Nature and the Right to Coerce in Kant's Doctrine of Right." *Archiv für Geschichte der Philosophie* 96(1), 126–139.
Ripstein, Arthur, 2009. *Force and Freedom. Kant's Legal and Political Philosophy*. Cambridge, MA: Harvard University Press.
Schönfeld, Martin. 2008. "The Green Kant: Environmental Dynamics and Sustainable Policies," in Louis P. Pojman and Paul Pojman, eds., *Environmental Ethics*. Boston: Thomson-Wadsworth.
Semple, Kirk, 2019. "Central American Farmers Head to the U.S., Fleeing Climate Change." *The New York Times*, April 13. https://www.nytimes.com/2019/04/13/world/americas/coffee-climate-change-migration.html
Timmermann, Jens. 2007. *Kant's Groundwork of the Metaphysics of Morals. A Commentary*. Cambridge: Cambridge University Press.
UNPFII (United Nations Permanent Forum on Indigenous Issues). 2019. "UN Meetings Coverage of the 18th Session of the UN Permanent Forum on Indigenous Issues, 20th Meeting." https://www.un.org/development/desa/indigenouspeoples/news/2019/05/un-meetings-coverage-of-18th-unpfii/
UNPFII (United Nations Permanent Forum for Indigenous Issues). no date. Factsheet. Who Are Indigenous Peoples? https://www.un.org/esa/socdev/unpfii/documents/5session_factsheet1.pdf

6

CONTRACTUALISM AND CLIMATE CHANGE

Paul Clements

Introduction

Over the 800,000 years prior to 1750 the concentration of carbon dioxide (CO_2) in the Earth's atmosphere ranged from about 180 to 300 parts per million (ppm), and average global temperatures ranged from 5°C below to nearly 1°C above the average for the last 10,000 years (Hansen and Sato 2011). The industrial revolution, however, fueled largely by fossilized plant remains, moved carbon from the earth to the atmosphere at increasing rates, and in 2018 (at this writing) the global average reached 412 ppm. This level of atmospheric CO_2, last observed 3 to 5 million years ago, corresponds to an equilibrium climate 2–3°C warmer than 2016 (World Meteorological Institution 2017, 1). The Intergovernmental Panel on Climate Change (IPCC) projects that by 2100 atmospheric CO_2 concentrations will lie between 550 and 1000 ppm, depending largely on human activities (Intergovernmental Panel on Climate Change 2014, Figure 1).

In the view of Sir David King, former chief scientific advisor to the UK government, potential harms from climate change constitute an existential threat to civilization (Taylor 2017).

While these harms are already significant, the physical and institutional momentum of global warming guarantee that they will increase dramatically. Notably, we probably cannot stop climate change from displacing perhaps 200 million people from their homes and livelihoods by 2050; the question is what happens after 2050. We are increasing atmospheric CO_2 much faster than ever before in Earth's history, and science cannot keep up with the profundity and reach of its consequences. The global community has agreed we must not let warming exceed 2°C above pre-industrial temperatures, but average global temperatures reached 1.3°C above pre-industrial levels in 2016 (ECMWF 2017), and plans for reducing carbon pollution are deeply inadequate. Also, harms predicted from 2°C of warming have been increasing.

At this writing the costly efforts needed to mitigate carbon pollution, to adapt to climate change, to support its victims, and to mediate resulting conflicts have been largely voluntary. The ethical approach known as "contractualism" focuses on fairness. It gives attention to victims and suggests that responsibility for climate change should be a significant factor in allocating costs. The fossil fuel industry, however, has effectively blocked reductions in carbon pollution. National governments are key for building the institutions to respond to climate change, but this has been impeded by our anarchic system of nation states. The extraordinary demands from climate change can only be met by stronger global governance. Contractualism offers a framework not only for sketching the outlines of a fair resolution of these demands, but also, allied with comparative political economy, for explaining the magnitude of the institution-building task, the resistance that must be overcome, and, given the structure of incentives, how solutions to climate change depend on moral heroes.

The Social Contract Tradition and Climate Change

The social contract tradition in ethical thought takes it that morality – what we owe to one another – is best understood in terms analogous to a contract. Being mutually dependent, humans have developed a sense of right, the foundation of morality, that is best understood in these terms. The two branches of the social contract tradition, however, contractualism and contractarianism, understand the construction of the sense of right differently.

For contractualism our sense of fairness, or right, is itself fundamental. Part of what it means to respect another person, we find, is to be able to justify our actions in terms he or she could freely accept. A task for philosophy is to articulate a framework that explains this act of justification and that can help to work out its consequences. Thus Scanlon argues that "an act is wrong if its performance under the circumstances would be disallowed by any set of principles for the general regulation of behavior that no one could reasonably reject as a basis for informed, unforced general agreement" (1998, 153). Clearly no one could reasonably reject that people should take responsibility when they hurt other people. For Scanlon this proposition has a simple explanation: to be human is to have a sense of right, which implies a sense of responsibility.

The other branch of the social contract tradition, contractarianism, discussed by Moehler in Chapter 7, takes self-interest to be fundamental and the sense of right to result from rational deliberation. Contractarianism "does not demand that agents share substantial moral ideals" (p. 141). It follows Hobbes (1651) in arguing that self-interested agents would contract to establish an authority that can impose rules for long-term, peaceful cooperation, or Gauthier (1986) in arguing that self-interest supports the adoption of moral principles.

The central ethical questions from climate change are about what responsibilities arise from the harms from carbon pollution, and, given that they involve stronger institutions of global governance, what principles these institutions should be based on and how they can be built. Like contractualism, contractarianism also finds a basis for a sense of responsibility, if only as a means to secure self-interest. For that matter, any legitimate ethical tradition must articulate a basis for the sense of responsibility, including utilitarianism, the ethical tradition that aims to maximize the good, with which social contract approaches are normally compared. Unless someone wants to argue that people who generate carbon pollution are not responsible for its consequences, responsibility must be a factor in building institutions of global governance for addressing climate change.

Once the momentum of climate change and the breadth and severity of its consequences are taken into account, any ethical approach that says people are responsible for the consequences of their actions will see that new institutions are needed for mitigation, adaptation, supporting victims, and mediating conflicts due to climate change. I don't know how a Hobbesian authority would approach it. A contractualist approach, most famously in Rawls' *A Theory of Justice* (1971), argues that to identify just principles for organizing society we should imagine that we do not know our place in society. In this "original position" we stand behind a "veil of ignorance," a point of view that excludes considerations that benefit us personally, and the same perspective applies to everyone. While Scanlon wants actions generally to be supported by principles that no one could reasonably reject, Rawls' veil of ignorance conditions reasonable rejection for principles of justice; both demand a consideration of fairness that is foreign to contractarianism. The original position situates persons conceived as free and equal such that unfair bargaining advantages and threats are ruled out (Rawls 2001, 15, 160).

In Rawls' "justice as fairness" equal liberties are fundamental, a society's economic inequalities should be in the long-term interest of its least well-off members, and the education system should support fair equality of opportunity (1971, 302). A fair approach to climate change involves not only rapid reductions in carbon pollution, but also particular efforts to reduce deaths and loss of livelihoods and to support global warming's victims.

Climate change is an externality from certain economic activities, an instance of the tragedy of the commons. When we burn coal, say to generate electricity, CO_2 is a by-product. Neither the seller nor the buyer of the electricity bears the cost of the carbon pollution. Rather it goes into the atmosphere where it may stay for 1000 years. Depending on how much additional carbon pollution is emitted, it contributes to warming the planet and to various consequences. The textbook economic response for a negative externality is to impose a tax equal to the cost of the harms. This forces seller and buyer to internalize the externality's cost and hence allows an efficient allocation of resources, if not a means to correct the harms. If the tax were set correctly, the level of carbon pollution would be economically rational. Perhaps with American leadership carbon taxes might have been imposed across industrialized countries after 1997, when the Kyoto Protocol, the first major climate change agreement to reduce emissions, was agreed, and CO_2 concentrations could have been kept below 400 ppm. But the US government refused to participate in a plan that did not include China and India. Carbon taxes could still be part of an efficient response to climate change, but by now a more proactive approach to reducing carbon emissions is needed.

Economic activities that produce CO_2 have been enormously profitable. Their greater beneficiaries tend to be politically powerful, while the victims of climate change, disproportionately in poorer countries and future generations, generally are not. Moreover, institutions needed to respond to climate change will increase global governance. While a fair approach gives equal consideration to the interests of beneficiaries and victims, this pits the weak against the powerful and it constitutes a threat to established institutions. The critical tasks involve building effective international institutions and covering their costs. But who shall bell the cat? All those in the relevant, privileged positions have other responsibilities if not vested interests in the status quo, and we may already have passed the point at which solutions exist that avoid deep moral conflicts.

Traditional principles of international justice endorsed by contractualism include a duty of non-intervention, and, as Rawls says, "a duty to assist other peoples living under unfavorable conditions that prevent their having a just or decent political or social regime" (1999, 37). Carbon pollution clearly abrogates the duty of non-intervention, but we cannot expect that the institutions needed to take responsibility for its harms will be freely adhered to by beneficiary countries' citizens. The likely consequences of failing to construct these institutions, however, are apocalyptic. Broadly, the more the task of building the needed institutions can be approached reasonably and rationally, and the sooner it can be carried out, the more harms from climate change can be reduced. Strategies based on superior power and delays tend to lead to more authoritarian approaches, toward the politics of the armed lifeboat. The alternative to building the required institutions, however, is a much darker version of the armed lifeboat.

Another contribution from a contractualist perspective is the idea that institutions are built from principles (Clements 2012). Climate change presents many challenges to the imagination. An important one is to see how principles that constitute existing institutions interact to generate current levels of carbon pollution, and how different institutional scenarios could plausibly keep the level of atmospheric CO_2 below 450 ppm (said to be the limit for keeping global warming to 2°C) and manage the harms from climate change. From today's perspective *any* solution (barring unforeseen scientific advances) is likely to appear utopian, but some are more plausible than others. Moving from one institutional framework to another involves not only constructing new norms or principles, but also overcoming resistance from those committed to principles that form the established framework. We have no metric for measuring the distance from one institution to another or for comparing the plausibility of two proposed institutional reforms, but comparative political economy offers methods for building the required imaginative and analytic capacities.

Consequences of Climate Change

While average global temperatures have exceeded pre-industrial levels by 1.3°C, over 90% of the excess heat trapped by greenhouse gas emissions has been absorbed into the oceans. If worldwide carbon emissions were halted, release of heat from the oceans would continue to raise atmospheric temperatures, perhaps by 0.6°C over a period of decades (Hansen 2005; Dahlman 2015). In addition, the economic life of existing global infrastructure commits us to a rise in CO_2 concentrations to roughly 450 ppm. Although the Paris Climate Agreement, adopted in December 2015, aims to hold the increase in the global average temperature to "well below 2°C above pre-industrial levels and to pursue efforts to limit the temperature increase to 1.5°C above pre-industrial levels" (United Nations 2015, Article 2:1a), given the physical and institutional momentum of global warming it appears unlikely these targets can be met.

In 2017 the United States saw $85 billion in costs in Texas and Louisiana from Hurricane Harvey, $45 billion in Florida from Hurricane Irma, and perhaps $95 billion in Puerto Rico from Hurricane Maria (Campoy 2017; Disis 2017), and over $100 billion in costs from wildfires in California (Lada 2017). Destruction from Hurricane Maria may cause more than a million Puerto Ricans to move to the United States (Cabranas and Lopez 2017). Starting in 2006 Syria experienced its worst drought in 900 years, forcing perhaps 1.5 million Syrians to move from farmlands to urban areas (Mansharamani 2016). This contributed to a civil war with about 5 million refugees. Climate change has also contributed to droughts across Africa, threatening millions with starvation (Gettleman 2017) and increasing migration to Europe (Sengupta 2016). Conflicting views on immigration within and among European countries threaten the very unity of the European Union (Laurence 2016; Schlein 2016; Rankin 2017).

Melting ice and thermal expansion of water are causing sea levels to rise. Roughly between 2013 and 2017 estimates of the likely rise by 2100 doubled, from a 1-meter to a 2-meter maximum, or even over 2.5 meters in an extreme scenario (Dahl 2017). Today there are about 60 million refugees and internally displaced people (IDPs) worldwide, already more than any time since World War II. One study reports that 2 feet of sea level rise by 2100 would submerge land home to about 100 million people, while more than 150 million would be displaced by 6 feet (Rutgers University 2017). Another study finds that 1 meter of sea level rise by 2060 could cause 1.4 billion people to become climate refugees, increasing to 2 billion (from a global population of 9–11 billion) with a 2-meter rise by 2100 (Geisler and Currens 2017, 323). Antarctica is the planet's largest repository of frozen water by far, but prior to 2014 the most authoritative estimates, from the IPCC, included less than a centimeter of sea level rise by 2100 from this source.

Some thought that Antarctica would contribute more than this, but the processes were not well enough understood to be included in official estimates. Now it turns out that besides Antarctic ice melting at the surface, when meltwater seeps into cracks its ice cliffs can collapse (hydrofracturing), a major contributor to the increased sea level rise estimates (Schlanger 2017). The last time atmospheric CO_2 was at current levels, about 3 million years ago, sea levels were 9 to 27 meters higher than today. Apparently we are only spared such effects today by "the time that it takes for the climate to respond and reach equilibrium with levels of atmospheric GHG [greenhouse gases]" (Environmental Justice Foundation 2017, 40).

It has long been known that global warming increases droughts both by reducing rainfall in some regions and by increasing evaporation, but it is hard to quantify the consequences. A 2018 study finds that aridification (substantial drying, increasing the likelihood of drought and forest fires and changing vegetation regimes) emerges over 24–32% of land surface when global temperatures rise 2°C above pre-industrial levels, but aridity is avoided in about two-thirds of this land if the temperature increase is kept to 1.5°C. Most of southern Europe is likely to become arid with 1.5°C warming, but Central and South America, southern Africa, coastal Australia, and southern China become mostly arid only as temperatures rise from 1.5 to 2°C (Park et al. 2018, 70–72).

Like rising sea levels, aridification and drought also displace large numbers of people, although these effects are even harder to quantify. People are also displaced by declining water flows in glacier-fed rivers, melting permafrost, and the loss of economic opportunity due to the death of coral and other sea life as oceans warm and become more acidic. Forest fires, inland floods, hurricanes, and heat waves strike with increasing frequency and intensity, displacing fewer people but leaving death and destruction in their wake.

Besides warming oceans, a quarter of our carbon pollution has been absorbed into the oceans, making them 30% more acidic than before the industrial revolution. Warmer and more acidic oceans make it harder for shellfish to build shells and reduce oxygen levels, harming many species at the low end of the food chain. Effects on micro-organisms, "which represent the vast majority of the biomass (organic carbon) in the ocean, will significantly alter the biogeochemical cycles and the functioning of the food webs at the global scale" (Danovaro et al. 2016, 57). But these and other effects of climate change on ocean chemistry are not well understood. There has also been no comprehensive study of the economic effects of the loss of fisheries due to the death of coral reefs.

In the 1950s the green revolution launched a global increase in agricultural productivity, but growth in yields has slowed for most major crops (Food and Agriculture Organization 2017, 48). Population growth and rising incomes are increasing food demand, and soil degradation and the mining of water supplies undermine agricultural productivity in many areas. Global warming increases agricultural productivity in some northern regions, but these effects are overtaken by factors that reduce yields as warming increases, and warming reduces agricultural productivity in the tropics. Around 1980 the continent of Africa tipped from producing a net surplus of food on average to a net deficit, and since then food imports have consistently increased. Climate change will deepen structural food deficits in Africa and push more regions into deficit. Whether global increases in agricultural production can keep up with global demand is an open question (Bourne 2015). Meanwhile, as the planet warms, weather patterns shift, and droughts and floods increase, temporary food deficits, both regionally and globally, will become more common. Simultaneous disruptions to agriculture in major "bread baskets" will likely lead to worldwide food shortages, which would be exacerbated by regional crises.

Contractualist Responsibility

Ethical questions about climate change arise at two levels. First, people in general and particularly people in advanced industrial countries bear responsibility for climate change, and some obligation to address both its causes, and, if they can, its consequences, but how do we understand these sensibilities? Of what do senses of responsibility and obligation consist, and how do we account for them? Second, what questions of social justice does climate change raise, and on what principles should they be resolved? What responsibilities does it place on nation states and other institutions, and what needs to be accomplished for these responsibilities to be fulfilled? And further, how do imperatives for institutions affect imperatives for individual persons?

A take on responsibility for climate change from Scanlon's contractualism might begin like this:

> Each of us is involved in systems of production and consumption, global in reach, that allow us to pursue the good as we see it. In these systems, we accept that activities and exchanges between persons and groups sometimes undermine opportunities for third parties. However, a principle that no one could reasonably reject as a basis for informed, unforced general agreement is that such activities and exchanges should not knowingly harm other persons. When they do knowingly impose harms, the parties that cause the harms bear responsibility for them. This responsibility can be discharged by fixing what is broken and removing the cause.

This is a bare-bones, inadequate conception of responsibility. It does not address thresholds, relations between individual and collective responsibility, or culpability and redress. To apply it to climate change and articulate specific duties to which it gives rise of course raises many questions. The point I would like to make is that a large part of moral theory is concerned with the possibility of and basis for a moral sense itself and hence for a sense of responsibility. While contractualism has its own approach, I think that for most people, the concept of responsibility sketched here is not controversial. If one takes account of climate science and the harms climate change is causing and likely to cause, it is clear that we have much to do. Indeed, I think an inkling of the weight of this responsibility contributes to many clinging to worldviews that exclude the science. The work for an ethics of climate change lies in articulating the responsibilities to which it gives rise, not the mere possibility of responsibility (although how one understands the latter may indeed influence the former).

International Justice

Given that it is increases in CO_2 that drive climate change and that the source can be anywhere on Earth, the starting point for questions of justice is the pathway of global CO_2 emissions, and each country's contributions. Higher CO_2 pathways benefit the current generation more (through greater energy production) and hurt future generations more, so the global pathway for CO_2 production begins as a question of justice between generations. Then, taking the nation state as the unit of responsibility, the question becomes how the global carbon budget is divided among countries. Under the Paris Agreement the division is based on voluntary national commitments and the "budget" is whatever actual emissions sum to, but this is inadequate. A related question is how the global budget is to be implemented. What international institutions are needed, such as to help countries reach their targets, for monitoring, and for sanctions for exceeding targets?

The harms climate change is inflicting will increase. More than 98% of deaths attributable to climate change occur in developing countries (DARA 2012, 23), and within countries the poor are generally more vulnerable. Wealth is correlated with historic and current energy production and hence responsibility for climate change, and vulnerability is correlated with poverty and living in the tropics, so at present and for the immediate future, harms from climate change tend to be inversely proportional to responsibility. The dangers from climate change are, however, somewhat predictable, and steps can be taken to reduce them, so responsibility for adaptation (adjusting to effects and reducing future harms) and for assisting victims of climate change should be considered together. Harms include not only immediate effects such as from droughts, storms, and floods, but also conflicts precipitated by competition for water and other affected resources and by forced population movements. Questions of justice include what should be done to adapt to climate change, to support victims, and to mediate conflicts, what new institutions are needed, and how costs should be allocated. As with steps to reduce CO_2 emissions, questions arise at international and national levels.

For Rawls, justice between generations involves moving toward and then sustaining a just society. Behind the veil of ignorance the parties in his original position do not know in which generation they find themselves, so they would select a savings principle such that, whatever their own time may be, all previous generations would have followed it (Rawls 2001, 160).

> Each generation must not only preserve the gains of culture and civilization, and maintain intact those just institutions that have been established, but it must also put aside in each period of time a suitable amount of real capital accumulation.
>
> *Rawls 1971, 285*

Once a just society is achieved, however, there is no obligation of justice to continue economic growth; real capital accumulation may fall to zero (Rawls 2001, 159).

But Rawls conceives of justice between generations in the context of principles of justice for a single people, the domestic arrangements of a nation state, and it is largely addressed by his just savings principle. As noted above, the principles relevant to climate change, discussed in his *Law of Peoples*, involve non-intervention and assistance. Rawls does not consider capital accumulation for one people undermining the gains of culture and civilization, or progress toward just institutions, for another, but his conception of justice would clearly condemn it.

Extreme weather events kill, but not randomly. Death is correlated with vulnerability and failure to prepare. Climate change undermines livelihoods and destroys habitats for segments of a society, sometimes gradually, such as through drought or rising sea levels, and sometimes in single or punctuated events, such as hurricanes or floods. Each country's polity consists of ongoing cooperation and competition carried out through established institutions and structures of authority. Loss of livelihood causes one to question institutions and authorities. If they are not in a position to secure it, as Hirschman's (1970) classic analysis explains, responses may be considered in terms of exit, loyalty, or voice. Parenti (2011) describes how climate change has contributed to violent conflicts, often with an ethnic dimension, in Kenya, Somalia, Kyrgyzstan, India, and Brazil, among other countries. Besides contributing to civil war in Syria, as noted above, drought is a significant factor in civil wars at this writing in Yemen and South Sudan and in the insurgency in northern Nigeria. Climate change is not, of course, solely responsible for these conflicts, but it will, most certainly, contribute to sparking and exacerbating more insurgencies and state violence.

Anthropogenic global warming sets natural cycles in motion that increase greenhouse gas concentrations in the atmosphere, accelerating warming. Methane is a shorter-lived but more potent greenhouse gas than CO_2. As warming melts permafrost in the Arctic, vegetation unfreezes at the bottom of ponds, and as it decomposes it produces methane. There are also large stores of frozen methane in sediments on the ocean floor. Some models predict a large-scale die-off of the Amazon rainforest before 3°C of warming, releasing significant carbon into the atmosphere. Like so many areas of climate science, these phenomena are not well enough understood for precise projections (Kopp et al. 2016, 352, 355–356). They indicate, however, that the current generation's responsibility for global warming could include responsibility for runaway increases even if anthropogenic carbon pollution is ended.

Given the duty of non-intervention, Rawlsian justice demands we keep CO_2 concentrations from exceeding 450 ppm and warming below 2°C. It is not that harms at this level are acceptable – rather a lower target is physically and institutionally implausible. In recent years the harms projected from 2°C of warming have risen, increasing the burden of responsibility on countries with higher carbon emissions. The distance between "bottom up" climate science, based on projecting forward observed trends, and "top down" scenarios, based on equilibrium conditions at 400 ppm CO_2 and above, has diminished somewhat. Today a reasonable bottom up scenario would involve, say by 2050, the number of IDPs and refugees rising from 60 to 260 million (possibly many more) due mainly to drought and rising sea levels. Many coastal cities are flooded and food shortages become more common and severe. Simultaneous weather disasters on the scale of Hurricane Maria hitting Puerto Rico become common. In this context we have to expect simmering conflicts to flare, governments to fall, and humanitarian institutions in many low- and middle-income countries to fail, so the fate of millions lies with international institutions. But these institutions are stressed not only by increased demands from crises, but also by political and economic repercussions from climate change within donor countries. Even with effective reductions in carbon emissions, ongoing carbon pollution and the physical momentum of global warming continue these harmful trends through the years to 2100 and beyond. And as CO_2 levels increase, warming continues, and science advances, we are likely to have more bad surprises.

In 2016 McKibben reported that, "to have a two-thirds chance of staying below a global increase of two degrees Celsius, we can release 800 gigatons more CO_2 into the atmosphere." This is, roughly, our carbon budget. To have a 50–50 chance of keeping the global temperature rise to 1.5°C, we could only release 353 more gigatons of CO_2, but operational coal mines and oil and gas wells worldwide in 2016 contained 942 gigatons worth of CO_2, and new wells continue to be dug (McKibben 2016). According to the IPCC, to have a more than even chance of staying below 2°C, global CO_2 emissions need to be lowered from 2010 levels by about 50% by 2050 and 90% by 2100 (Deep Carbon Pathways Project 2015, 3). Accounting for population growth, this implies a decline from 7 tons per person per year in 2010 to 2.2 tons in 2050 and 0.4 tons in 2100.

How should our carbon budget be divided? Total national commitments for carbon reductions in the Paris Agreement imply about 2.9°C of warming by 2100. But these commitments are voluntary, many countries are not on track to meet them, and they do not include consistent methods for calculating reductions, independent monitoring, or sanctions for failure to reach targets or for withdrawing from the agreement, as the United States has indicated it will do (Rogelj et al.

2016). After US rejection undermined the effectiveness of Kyoto, it proved difficult to come to a global agreement, and many saw Paris as a victory: the first international climate change agreement to reduce emissions that includes almost all nations of the world. But Paris commitments are clearly inadequate to secure Paris targets.

If most countries are voluntarily reducing emissions, each has an incentive to "free-ride." Given this incentive, an effective agreement needs not only commitments that reach the global target, but also independent monitoring and effective sanctions. From the perspective of social justice, the agreement to reduce CO_2 emissions must also address adaptation and harms. Countries that bear greater responsibility for climate change are generally wealthier and more politically powerful than more vulnerable countries, and more vulnerable countries have a greater incentive to come to an agreement. An agreement that favors the stronger parties, however, would exacerbate injustices from climate change, undermine the agreement's stability, and deepen resentments against the people of industrialized countries.

Rawls' original position offers a frame of reference for reaching a just agreement. When selecting principles of justice for a country, agents in the original position represent the country's citizens, but for the law of peoples Rawls' agents are representatives of each country. These are the parties who must come to a hypothetical consensus. I have argued elsewhere (Clements 2015) that due to the global reach of climate change and the unreliability of many governments in representing interests of their more vulnerable citizens, agents in an original position for climate change would represent individual persons anywhere in the world (as well as anywhere in time, starting with the present). Not knowing whom they represent, they would be particularly concerned to protect those whose lives climate change threatens. But in selecting principles for a climate change agreement they take account of plausibility, so they agree to a 2°C target.

Carbon emissions represent a scarce good; not knowing whom they represent, and appreciating the urgency of reductions, agents would identify equal per capita emissions for each country as the starting point or anchor for their agreement. They would, however, also find responsibility for climate change based on cumulative CO_2 emissions and ability to transition to renewable energy, both correlated with wealth, to be relevant. But these are also correlated with higher current CO_2, and hence the need for greater, more challenging reductions. The agents would find that countries with greater responsibility and capacity can discharge some of this responsibility by assisting other countries in their transitions (as well as with adaptation and supporting victims, as discussed below). They would find countries like the United States, for example, to have a greater "climate change obligation" than, say, India (Table 6.1), which could be satisfied in part by reaching the per capita target sooner and by assisting others.

National governments might employ a carbon tax or cap and trade system, along with proactive measures to increase energy efficiency and reduce carbon

TABLE 6.1 Greenhouse gas emissions (CO_2 equivalent), 2014.

	Total (gigatons)	*Percentage of world emissions*	*Per capita (tons)*	*GDP per capita (at purchasing power parity)*	*Cumulative CO_2 1850–2011*
World	49				
USA	6.3	14.4%	19.7	$54,657	27%
China	11.6	26.8%	8.5	$13,327	11%
India	3.2	6.6%	2.5	$5,797	3%

Sources: Total and per capita CO_2: Climate Watch (https://www.climatewatchdata.org/); GDP per capita: Comstat 2018; Cumulative CO_2: Ge et al. 2014. Cumulative figures only for CO_2, not including other greenhouse gases.

pollution. International carbon trades, however, where a firm in one country pays a firm in a second country to pollute less so the first firm can pollute more, raise concerns of justice. Even under private ownership, from the perspective of citizens a firm represents a national asset, and agents in the original position observe that neither the firm's owners nor the government can be relied upon effectively to represent the interests of the people. The agents would aim to balance efficiency gains from international carbon trading with defending the economic prospects of each nation's people. Of course many more details for a fair and workable mitigation agreement, such as for military activities, international transport, and to protect forests, need to be filled in.

Agents in the original position would find responsibility arising from cumulative emissions particularly relevant to the costs of adaptation, supporting victims, and mediating conflicts.

Given our anarchic international system this is the most challenging area of climate justice. Each government has an obligation of justice to its own people to protect them from harm and to assist when "weather" destroys their homes or livelihoods. But governments of poorer countries often lack not only the resources and institutions, but also, all too often, the political will to protect and support all their people. Rawls (1999, 5) distinguishes between "outlaw states" and "burdened societies," the former "regimes that refuse to comply with a reasonable Law of Peoples" (e.g., Myanmar, North Korea), the latter societies "whose historical, social and economic circumstances" make it difficult if not impossible for them to achieve a well-ordered regime (e.g., Afghanistan, Somalia).

With rich countries imposing bad weather on outlaw states and burdened societies, agents in the original position would find governments of countries with greater responsibility and capacity to be obligated to protect and support the likely and actual victims. Behind the veil of ignorance the imperative is clear: to prevent loss of life, protect and restore livelihoods, and build toward just institutions. There is a particular obligation to climate refugees – forced to leave their native homes due first to the failure to mitigate climate change, and then to the failure to protect or restore their livelihoods.

The obligation is to vulnerable and victimized individuals, not to their governments, which may be corrupt and authoritarian, favoring some ethnic groups or sections of society. It clearly conflicts with national sovereignty. But the obligation is also to support movement toward just institutions.

There is a certain fluidity in the content of obligations to mitigate climate change, to help with adaptation, to support victims, and to mediate conflicts. Faster mitigation yields less need for adaptation. More effective adaptation leads to fewer climate victims. Better support for victims in their home countries means fewer international climate refugees. And failures in all areas increase the likely contribution of climate change to domestic and international conflict. The obligation is greater where the impacts of climate change are greater and where institutions are weaker, and outlaw states and burdened societies present particular challenges.

What is the scale of the obligation to help with adaptation and to support victims? Consider that,

> Currently, forecasts vary from 25 million to 1 billion environmental migrants by 2050, moving either within their countries or across borders, on a permanent or temporary basis, with 200 million being the most widely cited estimate. … [S]ome 125 million people may be displaced by 2045 as a result of desertification.
>
> *Kamal 2017*

It costs about $3,000 a year to care for a refugee in Syria, and over $30,000 in, say, Germany or Austria (Williams 2016). For a first approximation, at $10,000 per person per year the financial cost to care for 100 million IDPs and refugees is $1,000 billion, while total economic and development assistance from industrial countries is around $150 billion a year at this writing.

The Political Economy of Climate Justice

Obligations for a country such as the United States include:

1. Adapt to climate change and support national victims
2. Reduce carbon pollution from 19 to 2 tons per person by or before 2050
3. Welcome and integrate a fair share of climate refugees, perhaps 5–10 million by 2050
4. Assist in covering costs for international mitigation, adaptation, supporting victims, and mediating conflicts
5. Assist in building institutions for international mitigation, adaptation, supporting victims, and mediating conflicts

At this writing the US is nowhere near a trajectory to fulfill these obligations, but failure to fulfill them increases the harm, the obligation, and the costs. At some point continuing failure to meet the international obligations (2–5 above) must render the US an outlaw state. Up to now this failure is largely due to concerted opposition from economic and ideological interests that effective action would threaten.

As early as 1978 James Black, a scientist at ExxonMobil, the largest, most profitable private corporation in the world, reported to company scientists and managers:

> [I]ndependent researchers estimated a doubling of the carbon dioxide (CO_2) concentration in the atmosphere would increase average global temperatures by 2 to 3 degrees Celsius (4 to 5 degrees Fahrenheit), and as much as 10 degrees Celsius (18 degrees Fahrenheit) at the poles. Rainfall might get heavier in some regions, and other places might turn to desert.
>
> *Banerjee et al. 2015, 1*

In the 1980s Exxon sought to be a good corporate citizen, sharing its scientific findings on the dangers of global warming. In 1989, however, it switched to protecting its financial interests, launching what turned into a campaign "to corrupt the debate on global warming through the funding of proxy groups that engage in denial and deception concerning climate science and through political lobby activities that involve the dissemination of disinformation and campaign contributions" (Kramer forthcoming). In the US presidential election campaign of 2000, candidate George W. Bush argued that serious action was needed to address global warming. As President, however, apparently due to pressures from oil interests inside and outside his administration, conservative think tanks, and conservative commentators such as Rush Limbaugh, he placed oil company executives in key regulatory positions and took only symbolic action to limit carbon pollution. A well-funded campaign followed the example from the tobacco industry to "manufacture doubt" about the scientific consensus on climate change and to purchase the support of key congresspeople to block climate change legislation (Kramer forthcoming).

We noted above that responses to the waves of refugees and other immigrants to the EU have threatened the EU's unity. Immigration to the United States has provoked similar nativist reactions. These contributed to the election of Donald Trump to the presidency on a platform including promises to expel millions of immigrants who entered the country illegally and to build a wall along the border with Mexico.

Rising sea levels and storm-driven flooding probably present the greatest direct economic threat to the US from climate change, as they undermine the value of coastal properties. The US federal government, however, still subsidizes insurance on these properties. Even though the agency that covers these policies is over \$24 billion in debt, efforts to remove the subsidies have been obstructed by congresspeople representing coastal districts. Congress has also failed to have floodplain maps updated so homeowners and buyers would better understand the risks they face (*New York Times* Editorial Board 2017); a plausible collapse in America's waterfront property market could have an economic impact greater than the bursting of the real estate bubble that precipitated the 2008 US recession (Urbina 2016).

Besides these domestic institutional failures, the absence of American leadership has also impeded the establishment of effective international climate change institutions. We have noted that voluntary national commitments in the Paris Agreement, many of which are not being kept, would lead to about 3°C of warming. Governments made a separate financial agreement to provide \$100 billion annually for mitigation, adaptation, and supporting victims, but institutional arrangements are vague, nor is it clear how far this inadequate sum will be additional to current aid. President Trump's withdrawing the US from the Paris Agreement and cutting foreign aid commitments undermine even these weak agreements.

Recall that agents in the original position would be particularly concerned to protect those whose lives and livelihoods are threatened by climate change, and since each government also has this obligation to its own people, the obligation for countries with more responsibility is greater where government failure is more severe. Today's relief and refugee support organizations, such as the World Food Programme, the UN High Commissioner for Refugees, and the host of NGOs that assist disaster victims and refugees are struggling to cope with current needs, and support is often inadequate. Support for refugees and IDPs depends on official arrangements with governments, but some governments are not reliable representatives of victims' interests. The government of conflict-ridden South Sudan, for example, impedes support for many of its people. As numbers of IDPs and refugees double and triple, simmering conflicts will be exacerbated and government failures in their humanitarian obligations will increase. Drought, declining water supplies, and other degradation of natural resources influenced by climate change will increase conflicts over such resources within and between states. In some such cases the obligation to support victims will conflict directly with state sovereignty, often in a context of violence.

The international community has always had a humanitarian obligation to support victims of war and natural disaster. When climate change contributes to the causes, humanitarian obligation is supplanted by direct responsibility. When state authorities in affected countries fail to facilitate support for victims, such as when they are attacking them in a civil war, engaged in conflict with another state, or simply neglecting their needs, this responsibility may entail taking control over territory and/or institutions against the will of the sovereign (or the warring parties). Then this territory and/or these institutions must be managed and protected, and just institutions (e.g. of self-determination) established in a potentially hostile environment with uncongenial patterns of authority. Yet climate change has already stretched humanitarian institutions to their limits, humanitarian crises will multiply, and crises in burdened and outlaw states are likely sometimes to coincide with crises in advanced industrial countries.

Responsibility for climate change lies with individual states, but states have always used foreign aid and development assistance in part as an instrument to pursue their political interests. "[T]he United States has viewed all multilateral organizations, including the World Bank, as instruments of foreign policy to be used in support of specific U.S. aims and objectives" (Gwin 1997, 195). Control over territory exacerbates this political risk. The only way the allocation of the substantial resources needed for mitigation, adaptation, and supporting victims, and particularly for armed intervention and control of territory, can be protected from the narrow interests of donor countries is to establish firm barriers against national influence. High management standards and accountability are also critical. Not only the independence and resources but also the management capacities of international institutions need to be greatly increased.

The act of building institutions includes overcoming resistance from those whom the new normative order threatens. Costs to a nation such as the US for national and international efforts require cuts elsewhere in the federal budget and/or increased taxes. Coal, petroleum, and natural gas interests will continue to work to maintain the value of their assets. The greater threats to established principles, however, probably involve welcoming and integrating climate refugees and building international institutions to which countries must contribute but which are insulated from national political influence.

National financial obligations and obligations to build international institutions have to be recognized before they can be fulfilled. American political experience suggests that it will take international crises with much loss of life before the legislative and executive branches will come to grips with this question. In such events the alternatives are likely to lie on a spectrum between nativist/nationalistic with militarized responses and cooperative and internationalist. But a nationalistic, militarized response can only lead to worldwide hegemony, or to competing armed lifeboats, and it is unlikely that either could contain worldwide carbon pollution.

Some libertarian individualists, business interests, conservative Christians, nativists, and no doubt others are likely to resist financial obligations, specific mitigation and adaptation programs, and influxes of climate refugees. Federal programs will variously confront individualistic responses and assertions of local and state powers. But the federal government is merely the agent for fulfilling national obligations to climate victims such as the US, or any other country for that matter, has never before experienced; obligations that apply to all residents. It will therefore fall to the federal government to impose obligations, often breaking with precedent. Resistance will follow, in light of established political culture, inside and outside the law, including resistance against climate refugees, and the federal government, along with loyal local and state authorities, will have to quell or transform it. They will have to defend the security and interests of refugees against various attacks. Since directives will sometimes originate from officials in international institutions, the federal government will rightly be accused of compromising national sovereignty. The source of this compromise, however, is the harm imposed on others by carbon pollution from the national territory.

Stability in Conceptions of Justice and the Instability of Climate Justice

At this writing, acceptance of responsibility for the harms from climate change has been a voluntary choice of persons or institutions (including states). Decisions to mitigate, to adapt, and to support victims have been, in an important sense, voluntary. As current and expected harms increase, however, free-riding will become increasingly unacceptable, and justice demands that responsibility should be a significant factor in allocating costs. Agents who do not accept responsibility will experience the imposition of costs as a violation, and indeed, absent responsibility, it would often be a violation of individual rights, principles of self-determination, and state sovereignty. And while financial costs can be divided mathematically, the movement of persons who lose their homes or livelihoods creates a variety of responsibilities that necessarily fall more heavily on some than on others. It falls to governments of advanced states to impose and coordinate costs and obligations within their territories, and, given the threat of free-riders in an anarchic world, supranational institutions are needed to impose and coordinate costs and obligations on states. Stronger supranational institutions are also needed to support victims in poor countries, sometimes in violation of national sovereignty.

The longer the delay in building these institutions, the worse the harms and the higher the costs. In particular, the sooner mitigation and adaptation are accelerated, the fewer IDPs and refugees climate change will create, and the less conflict it will generate. Opponents of reform, however, have so far been very effective. While they will abandon some strategies as the state of play evolves, they can be expected to launch new ones, and new opponents will arise as costs and responsibilities increase.

A distinctive feature of justice as fairness is its focus on the long-term interests of the least well-off members of society, associated with the veil of ignorance. It is this veil that leads us to emphasize lives and livelihoods. Agents in the original position do not know if they represent a banker in New York City or a farmer no longer able to eke out a living in a drought-stricken Sahel. Given that the banker's wealth is due in part to America's cumulative carbon pollution, the agents would find the banker to have a positive obligation to protect the farmer's life and livelihood, mediated by (among others) the American government, possibly including helping to settle the farmer in New York. Notably, due to the banker's responsibility for the farmer's loss, this is a stronger obligation than that to provide humanitarian assistance. Notably also, this obligation generates incentives that can undermine the self-reliance of persons and governments, as some may pretend they are less capable than they actually are, or even become less capable, in order to increase eligibility for support.

The strong obligations of the state to the least well-off in justice as fairness, such as to support a living income, opportunities for employment, and good education, have led some to question the stability of Rawls' conception of justice. Can wealthy citizens' sense of justice bring them to support laws that increase their own taxes and undermine their political power? Or is it more likely that they may:

> *Like to be* less materialistic than they are – indeed they would like to be less materialistic because they recognize that the principles of justice do not impose unreasonable burdens on them — but they remain powerfully tempted to acquire more than justice allows.
>
> *Weithman 2015, 108*

If it turns out that human psychology makes a conception of justice inherently unstable, the appeal of such a conception is significantly diminished (Weithman 2015). Weithman argues that:

> Contractualism itself is underpinned by a kind of moral faith that principles can be freely adhered to by those who are subject to them. Crudely put, this faith is faith in the amenability of human beings to the relevant kind of moral education and in the ability of basic institutions to deliver it.
>
> *Weithman 2015, 93*

He suggests that contractualists are likely to believe that this faith is reasonable within a national community, but not to support equivalent duties of justice to the global poor (2015, 133).

The duties of citizens of rich countries to victims of climate change among the global poor are qualitatively different from duties of justice that wealthy citizens may owe the poor in their own national community. The former arise from more immediate and direct responsibility for this victimization, due partly to conditions for their wealth, and, particularly in light of the momentum of climate change and the severity of harms, these duties are stronger. While contractualism represents one family of approaches to social justice, duties to climate victims stem from responsibility, which is fundamental to any moral theory.

The recognition of this responsibility leads to duties to build institutions that strengthen global governance, undermine the sovereignty of nations such as by taking control of territory, impose significant financial obligations, and require states to impose not only financial costs on their citizens but also measures to welcome and integrate millions of refugees, most of whom are at a great cultural distance from natives, and, of course, who did not leave their homes simply by choice. To strengthen global governance is to impose new rules on each level of government down to the individual, diminishing liberties, powers, and resources. There is no purpose more fundamental to democracy than to support individual liberty and autonomy, and, of course, in light of climate change, this is precisely what stronger global governance accomplishes, but the necessary "reshuffling" of norms or principles most immediately diminishes many freedoms and threatens established orders. Yet when threatened, individuals and institutions often fight back.

Actions needed to respond to climate change not only threaten many people's interests but also their principles and worldviews. Heretofore many have responded to this threat, including the current President of the United States and Director of the Environmental Protection Agency, simply by denying reality. To sustain democracy, governments first need to acknowledge responsibility democratically. To do so is costly, but delay increases costs exponentially. Democratic methods for getting democratic institutions to acknowledge their principles include elections, lawsuits, new law, civil disobedience, and other actions that dramatize the injustice. A strong coalition of countries participating in fair global climate governance could leverage participation by free-riders. When injustice causes suffering and death, however, responses have often exceeded liberal boundaries, such as with attacks on persons and institutions viewed as perpetrators.

The imperative, therefore, is to forge a democratic consensus for institutions of global governance that can move us from voluntary and decentralized compliance with climate justice – which has proved inadequate – to enforced compliance based on responsibility. This move is, however, massively destabilizing. It threatens powerful interests in order to avoid harms to agents who, for the most part, cannot defend their own interests. It therefore falls to self-selected agents who are prepared to undertake the costs of action on behalf of climate victims and the common good.

Future harms from a given level of carbon pollution massively exceed current harms, and even current harms are so many and widespread, with so many interactions, that their magnitude is hard to grasp. Based on the geological record for concentrations of atmospheric CO_2 and the most favorable trajectories, civilization is in mortal peril. Bottom up climate science has greatly underestimated the harms. A current summation is far more terrible than one from just five years ago, and there are sure to be more bad surprises. The consequences of failure to take extraordinarily rapid action to build the required institutions are far greater than those from failure to address other environmental or social challenges such as fracking, plastic in the ocean, opioid abuse, sexual harassment, or gun control.

Only national executives and legislatures can initiate the institutions of global governance required for mitigation, adaptation, supporting victims, and mediating conflicts driven by climate change. In order to sustain democratic institutions as the planet warms, the priority is to elect executives and legislators who support this agenda. It is clear that the scale of each of these challenges – mitigation, adaptation, supporting victims, and mediating conflicts – is such that it can only be managed with much stronger global institutions. We need these institutions today, although of course it takes time to build them. It is also clear that decentralized individual and local actions, to reduce carbon footprints, prepare for floods and heat waves, care for refugees, and mediate conflicts, are inadequate. Progress in institution building can be made before global consensus is achieved on the full program of global governance, and while the full program is needed today, elements can be achieved incrementally. Delay merely increases harms.

Bibliography

Bannerjee, Neela, Lisa Song, and David Hasemyer. 2015. "Exxon's Own Research Confirmed Fossil Fuels' Role in Global Warming Decades Ago." *Inside Climate News*, September 15, 2015. https://insideclimatenews.org/news/15092015/Exxons-own-research-confirmed-fossil-fuels-role-in-global-warming.

Bourne, Jason. 2015. *The End of Plenty: The Race to Feed a Crowded World*. New York: Norton.

Cabranas, Jose and Felix Lopez. 2017. "The Puerto Ricans are Coming." *The Washington Post*, September 27, 2017. https://www.washingtonpost.com/opinions/the-puerto-ricans-are-coming/2017/09/27/aa49a5fa-a3b6-11e7-8cfe-d5b912fabc99_story.html?utm_term=.556a7bacf08d.

Campoy, Ana. 2017. "How Hurricane Maria Will be More Costly for Puerto Rico than Harvey was for Texas." *Quartz*, September 27, 2017. https://qz.com/1088762/puerto-rico-hurricane-marias-devastating-economic-cost/.

Clements, Paul. 2015. "Rawlsian Ethics of Climate Change." *Critical Criminology* 23 (4): 461–471.

———. 2012. *Rawlsian Political Analysis: Rethinking the Microfoundations of Social Science*. Notre Dame, IN: University of Notre Dame Press.

Comstat. 2018. "GDP per Capita by Country; Statistics from IMF, 1980–2022." http://comstat.comesa.int/pjeqzh/gdp-per-capita-by-country-statistics-from-imf-1980–2022?country=India.

Dahl, Kristy. 2017. "Fast and Getting Faster: The Verdict on Sea Level Rise from the Latest National Climate Assessment." [Blog] Union of Concerned Scientists, November 3, 2017. https://blog.ucsusa.org/kristy-dahl/nca-sea-level-rise.

Dahlman, LuAnn. 2015. *Climate Change: Ocean Heat Content*. NOAA Climate.gov. https://www.climate.gov/news-features/understanding-climate/climate-change-ocean-heat-content.

Danovaro, Roberto, Cinzia Corinaldesi, Eugeion Rastelli and Antonio Dell'Anno. 2016. "Impacts and Effects of Ocean Warming on Micro-Organisms." In *Explaining Ocean Warming: Causes, Scale, Effects and Consequences*, edited by Daniel Laffoley and J. M. Baxter, 57–75. Gland, Switzerland: IUCN.

DARA and the Climate Vulnerable Forum. 2012. *Climate Vulnerability Monitor.* 2nd ed. *A Guide to the Cold Calculus of a Hot Planet*. Madrid, Spain: Fundacion DARA Internacional.

Deep Decarbonization Pathways Project. 2015. *Pathways to Deep Decarbonization: 2015 Report*. SDSN-IDDRI.

Disis, Jill. 2017. "Hurricane Maria Could be a $95 Billion Storm for Puerto Rico." *CNN Money*, September 28, 2017.

ECMWF. 2017. "2016 Was the Warmest Year Yet, ECMWF Data Show." ECMWF. January 5, 2017. https://www.ecmwf.int/en/about/media-centre/news/2017/2016-was-warmest-year-yet-ecmwf-data-show.

Environmental Justice Foundation. 2017. *Beyond Borders: Our Changing Climate – Its Role in Conflict and Displacement*. London: Environmental Justice Foundation.

Food and Agriculture Organization. 2017. *The Future of Food and Agriculture: Trends and Challenges*. Rome, Italy: FAO.

Gauthier, David. 1986. *Morals by Agreement*. Oxford: Clarendon Press.

Ge, Mengpin, Johannes Friedrich, and Thomas Damassa. 2014. "6 Graphs Explain the World's Top 10 Emitters." World Resources Institute. November 25, 2014. https://wri.org/blog/2014/11/6-graphs-explain-world%E2%80%99s-top-10-emitters.

Geisler, Charles and Ben Currens. 2017. "Impediments to Inland Resettlement Under Conditions of Accelerated Sea Level Rise." *Land Use Policy* 66: 322–330.

Gettleman, Jeffrey. 2017. "Drought and War Heighten Threat of Not just 1 Famine, but 4." *The New York Times*, March 27, 2017. https://www.nytimes.com/2017/03/27/world/africa/famine-somalia-nigeria-south-sudan-yemen-water.html.

Gwin, Catherine. 1997. "U.S. Relations with the World Bank, 1945–1992." In *The World Bank: Its First Half Century, Volume Two: Perspectives*, edited by Devesh Kapur and John P. Lewis, 195–274. Washington, DC: Brookings Institution Press.

Hansen, James and Makiko Sato. 2011. "Earth's Climate History: Implications for Tomorrow." *Science Briefs*. National Aeronautics and Space Administration: Goddard Institute for Space Studies. https://www.giss.nasa.gov/research/briefs/hansen_15/.

Hansen, James. 2005. "Earth's Energy Imbalance: Confirmation and Implications." *Science* 308 (5727): 1431–1435.

Hirschman, Albert. 1970. *Exit, Voice and Loyalty*. Cambridge, MA: Harvard University Press.

Hobbes, Thomas. 1651. *Leviathan*. Edited by Richard Tuck. 1996 ed. Cambridge: Cambridge University Press.

Intergovernmental Panel on Climate Change. 2014. "Carbon Dioxide: Projected Emissions and Concentrations." Data Distribution Center, last modified April 4, 2014. www.ipcc-data.org/observ/ddc_co2.html.

Kamal, Baher. 2017. "Climate Migrants Might Reach One Billion by 2050." Reliefweb, August 21, 2017. https://reliefweb.int/report/world/climate-migrants-might-reach-one-billion-2050.

Kopp, Robert E., Rachael L. Schom, Gernot Wagner, and Jiacan Yuan. 2016. "Tipping Elements and Climate–Economic Shocks: Pathways toward Integrated Assessment." *Earth's Future*, 4: 346–372.

Kramer, Ronald C. forthcoming. Carbon Criminals, Climate Crimes. New Brunswick, NJ: Rutgers University Press.

Lada, Brian. 2017. "Devastating California Wildfires Predicted to Cost US Economy $85 Billion; Containment may Take Weeks." *Fox News*, October 15, 2017. https://www.accuweather.com/en/weather-news/devastating-california-wildfires-predicted-to-cost-us-economy-85-billion-containment-may-take-weeks/70003000.

Laurence, Peter. 2016. "How is the Migrant Crisis Dividing EU Countries?" *BBC News*, March 4, 2016. www.bbc.com/news/world-europe-34278886.

Mansharamani, Vikram. 2016. "A Major Contributor to the Syrian Conflict? Climate Change." *PBS News Hour*, March 17, 2016. https://www.pbs.org/newshour/economy/a-major-contributor-to-the-syrian-conflict-climate-change.

McKibben, Bill. 2016. "Recalculating the Climate Math: The Numbers on Global Warming are Even Scarier than We Thought." *New Republic,* September 22, 2016. https://newrepublic.com/article/136987/recalculating-climate-math.

New York Times Editorial Board. 2017. "How Federal Flood Insurance Puts Homes at Risk." *The New York Times*, August 31, 2017. https://www.nytimes.com/2017/08/31/opinion/flood-insurance-program-.html.

Parenti, Christian. 2011. *Tropic of Chaos and the New Geography of Violence*. New York: Nation Books.

Park, Chang Eui. 2018. "Keeping Global Warming Within 1.5°C Constrains Emergence of Aridification." *Nature Climate Change* 8: 70–74.

Rankin, Jennifer. 2017. "Bitter Divisions over Migration Threaten Show of Unity at EU Summit." *The Guardian*, December 14, 2017. https://www.theguardian.com/world/2017/dec/14/divisions-over-migration-spoil-show-of-unity-at-eu-summit-in-brussels.

Rawls, John. 2001. *Justice as Fairness: A Restatement*. Edited by Erin Kelly. Cambridge, MA: Harvard University Press.

———. 1999. *The Law of Peoples: with "The Idea of Public Reason Revisited*. Cambridge, MA: Harvard University Press.

———. 1971. *A Theory of Justice*. Cambridge, MA: Harvard University Press.

Rogeli, Joeri. 2016. "Paris Agreement Climate Proposals Need a Boost to Keep Warming Well Below 2°C." *Nature* 534: 631–639.

Rutgers University. "Sea-Level Rise Projections made Hazy by Antarctic Instability." Phys.org. December 13, 2017. https://phys.org/news/2017-12-sea-level-hazy-antarctic-instability.html.

Scanlon, Thomas M. 1998. *What We Owe to Each Other*. Cambridge, MA: Harvard University Press.

Schlanger, Zoe. 2017. "Land Where 153 Million People Live may be Underwater by 2100." *Quartz*, December 16, 2017. https://qz.com/1158601/over-153-million-homes-will-be-underwater-by-2100/.

Schlein, Lisa. 2016. "Refugee Crisis Threatens European Unity, UN Official Says." *VOA News*, February 5, 2016. https://www.voanews.com/a/refugee-crisis-european-unity-united-nations-official-peter-sutherland/3178836.html.

Sengupta, Somini. 2016. "Heat, Hunger and War Force Africans onto a 'Road on Fire'." *The New York Times,* December 15, 2016. https://www.nytimes.com/interactive/2016/12/15/world/africa/agadez-climate-change.html.

Taylor, Matthew. 2017. "Climate Change 'Will Create World's Biggest Refugee Crisis'." *The Guardian*, November 2, 2017. https://www.theguardian.com/environment/2017/nov/02/climate-change-will-create-worlds-biggest-refugee-crisis.

United Nations Framework Convention on Climate Change. 2015. *Paris Agreement*. United Nations. https://unfccc.int/files/meetings/paris_nov_2015/application/pdf/paris_agreement_english_.pdf.

Urbina, Ian. 2016. "Perils of Climate Change could swamp local Real Estate." *New York Times*, November 24, 2016. https://www.nytimes.com/2016/11/24/science/global-warming-coastal-real-estate.html.

Weithman, Paul. 2015. "Relational Equality, Inherent Stability and the Reach of Contractualism." *Social Philosophy & Policy* 31 (2): 92–113.

Williams, Robin. 2016. "Syrian Refugees Will Cost Ten Times More to Care for in Europe than in Neighboring Countries." *Independent*, March 13, 2016. https://www.independent.co.uk/voices/syrian-refugees-will-cost-ten-times-more-to-care-for-in-europe-than-in-neighboring-countries-a6928676.html.

World Meteorological Organization. 2017. "The State of Greenhouse Gases in the Atmosphere based on Global Observations through 2016." *WMO Greenhouse Gas Bulletin* 13: 7. https://ane4bf-datap1.s3-eu-west-1.amazonaws.com/wmocms/s3fs-public/ckeditor/files/GHG_Bulletin_13_EN_final_1_1.pdf?LGJNmHpwKkEG2Qw4mEQjdm6bWxgWAJHa.

7

CONTRACTARIANISM AND CLIMATE CHANGE

Michael Moehler

Introduction

The ethics of climate change entails local and global dimensions and, although these two dimensions cannot always be separated clearly, I focus primarily on the global dimensions of climate change. The main reason for this focus is that, although societies may agree on local measures that help mitigate the effects of human causes of climate change and are enforced by local institutions, some of the most significant human causes of climate change, such as the emission of greenhouse gases into the atmosphere (in particular carbon dioxide, but also short-lived climate pollutants, such as methane, black carbon, and fluorinated gases),[1] represent problems of collective action with an inherently global scope. The warming of our planet does not stop at national borders and is, in its cause and effect, globally interdependent. As such, even if some societies agree on measures that help to reduce global warming and enforce such measures locally, if other societies reject these measures or they reach, based upon their particular moral views, different conclusions about the adequate measures to address climate change, then such local efforts, although not entirely in vain, will often be inefficient and ineffectual.

Methodologically, three core assumptions underlie my discussion. First, I assume that non-natural causes of climate change, in particular the emission of greenhouse gases, exist and negatively affect human wellbeing and ultimately human existence. That is, I assume that natural processes, such as changes in the sun's energy, do not fully explain the currently observed level of global warming but that human activity contributes to it. Anthropogenic climate change occurs.[2] Second, I assume deep moral pluralism with regard to questions of climate change. I assume that agents may hold different, irreconcilable, or no moral views at all concerning questions of climate change. Third, considering the far-reaching implications of climate change, I consider it to be essential to employ a moral framework that (i) is sensitive to empirical facts, (ii) justifies moral conclusions that can be implemented, and (iii) applies globally despite deep moral pluralism.

One moral framework that fulfills these conditions is "moral contractarianism." This approach goes back to Hobbes's moral theory (Hobbes 1651) and has been defended more recently by Gauthier (1986), Kavka (1986), and me (Moehler 2018, 2020), among others. However, under the circumstances described, especially the global dimension of some of the most significant human causes of climate change, it is unclear whether this approach justifies any specific moral demands concerning questions of climate change in our current world. In the following, I argue that moral contractarianism does justify such demands and that the approach can be used to guide our considerations concerning questions of climate change.

Nevertheless, I am not concerned primarily with determining the precise content of the moral demands of climate change. Instead, as is not uncommon in the tradition of moral contractarianism, I assume that the actual agreements that agents reach on questions of climate change determine the content of such demands. In particular, for my argument I rely on the Paris Agreement because it is the most recent climate change agreement. The main focus of my discussion lies on the fundamental problems associated with Hobbesian moral contractarianism that seem to threaten the implementation of any agreement that is reached by this approach in our current world. If applied to the global level, Hobbes's moral theory suggests that it is irrational to comply with agreements reached by this approach without the institution of a global Leviathan, or at least a world court and world police (Moehler 2009), that could enforce the agreements. However, such global institutions in Hobbes's strict sense currently do not exist in our world.

Despite this feature of Hobbes's theory, I suggest a "Hobbesian" solution to the problem of climate change that does not assume a global Leviathan but relies on the distinction between the concepts of "overlapping consensus" and "*modus vivendi*." An overlapping consensus assumes agreement among agents on moral standards that are based on substantial moral grounds whereas a *modus vivendi* assumes agreement merely on instrumental grounds and a common goal among agents. In a deeply morally pluralistic world, an overlapping consensus on the moral demands of climate change is typically not feasible. Nevertheless, I argue that if agents are prudent, then a *modus vivendi* may be sufficient as a starting point to address some of the most pressing issues of climate change. Specifically, I argue that the shift in climate change negotiations from the Kyoto Protocol to the Paris Agreement could be considered to be guided by reasoning that underlies Hobbesian moral contractarianism, and thus, this Hobbesian solution to climate change may be close to implementation.

Hobbes's Moral Contractarianism

In contemporary philosophy, the position of moral contractarianism originates with Hobbes's moral theory.[3] Moral contractarianism justifies moral standards based upon agreement among agents who are affected by these standards. The central idea of moral contractarianism is that, if agents agree to the moral standards that govern their interactions, then, all things considered, the agents have no reason to reject the authority of these standards because the agents themselves have authorized the standards. Moral contractarianism respects the autonomy of agents to approve the moral standards that apply to them, and this feature of moral contractarianism ensures that the moral standards justified are considered to be mutually beneficial.

More specifically, the position of moral contractarianism assumes that agents are (i) rational in that they pursue predominantly their own interests and (ii) roughly equal by nature in that they pose a potential threat to each other. According to Hobbes (1651, Part 1, Chapter 13), the weakest is able to kill the strongest. Moral contractarianism does not demand that agents share substantial moral ideals, or hold such ideals at all, as a starting point for the justification of moral standards (although it does not rule out that agents hold such ideals). Instead, the approach assumes merely that agents, despite their different starting points, share a common end, which, in Hobbes's case, is the goal of ensuring peaceful long-term cooperation. The position of moral contractarianism relies on means-end reasoning and, from a moral perspective, does not assume a substantial moral basis among agents for the justification of moral standards.

Hobbes's social contract theory is based on a hypothetical state of nature that assumes, apart from rough natural equality among agents, that agents pursue their own good and are free to do what they consider necessary to preserve their lives, which includes taking the lives of others if doing so is necessary to preserve their own lives (Hobbes 1651, Part 1, Chapter 14). In the state of nature, agents have threat capacities against each other and no binding contractual moral obligations. Stated positively, Hobbes models the state of nature as a situation of deep moral pluralism in which agents may hold any moral ideals, or no moral ideals at all. This characteristic of Hobbes's state of nature, together with the assumption of agents' competition, diffidence, glory-seeking, and desire for power in an environment of scarce resources, leads to a situation of severe conflict in which agents recognize that striking first in their struggle for life may provide them with the necessary advantage to survive. Such preemptive action makes the state of nature a war of all against all that destroys almost all productive efforts, which conflicts with agents' desire for self-preservation (Hobbes 1651, Part 1, Chapter 13).

In addition to preserving their lives, Hobbes (1651, Part 1, Chapters 10 and 11) assumes that agents are prudent in that they have an interest in securing their long-term wellbeing. Prudent agents aim to secure the fulfillment of their interests not only today, but also in the future. According to Hobbes, prudence dictates to rational agents the laws of nature that demand that agents do not perform actions that are destructive to their lives or take away the means to preserve them. The first law of nature requires agents to seek peace because, under the circumstances described, rational agents expect that peaceful long-term cooperation is most beneficial to them. According to Hobbes, the formation of society is a necessary means to reach this end. The second law of nature states that society can be established only if all agents lay down their rights of nature and transfer these rights to a common authority that is not part of the society to be formed. According to Hobbes, only if all agents give up their rights to govern themselves and transfer these rights to an external authority with unlimited and undivided power can society be established and maintained. The third law of nature demands that agents keep their contracts once these contracts are binding. That is, once society is established, agents must honor their agreements in order for peaceful long-term cooperation to be ensured.

Hobbes recognizes that two problems of collective action must be solved in order for rational agents to leave the state of nature and maintain peace: the problem of assurance and the problem of compliance. Concerning the former, if agents keep their rights of nature, then they will remain in the state of nature and face a war of all against all, which is suboptimal for everyone. In the state of nature, agents cooperate with each other only if they can be sure that other agents will also cooperate by transferring their rights of nature to an external authority that is strong enough to enforce any subsequent agreements. According to Hobbes (1651, Part 2, Chapter 17), only when a common power is in place that is strong enough to enforce the agreements that have been made are those agreements binding and a valid social contract is established. According to Hobbes, only an external sovereign can solve the problem of assurance in the state of nature by replacing interpersonal trust, or lack thereof, by institutional trust. Hobbes's solution to the problem of assurance is captured by his second law of nature.

Once society is formed, the problem of compliance arises. Agents' agreements represent the social contract and the origin of justice that, according to Hobbes, demand simply to keep one's agreements in order to ensure peaceful long-term cooperation. The problem of compliance is expressed by Hobbes's discussion of the 'Foole'.[4] The Foole recognizes that, in the long term, being part of a functioning society is more beneficial than staying in the state of nature. In the short term, however, the Foole defects from moral standards whenever doing so is advantageous. The Foole, like many other members of society, aims to satisfy primarily her own interests and decides the best behavior for herself in each instance. However, according to Hobbes, the Foole is shortsighted.[5] The Foole does not fully consider the negative effects of her defecting behavior or assumes that such behavior will remain undetected. The Foole believes that she can have both the benefits of peaceful long-term cooperation and the gains from exploiting others in the short term.

In order to prevent such free-riding once society is established, the sovereign must institute a police system that ensures a reasonably high apprehension and conviction rate, assuming that the costs for instituting such a system are lower than the gains from peaceful long-term cooperation. In addition, because in practice it is usually not possible to implement an enforcement system that detects all free-riders, the sovereign must threaten the members of society with severe sanctions for defective behavior. According to Hobbes (1651, Part 1, Chapter 15), the sovereign may even exclude free-riders from cooperation in general or take their lives in order to ensure peace. If such severe sanctions are in place, then free-riding generally does not pay off. In this case, even opportunistic case-by-case decision makers will follow the established rules because, if free-riding is likely to be detected and the sanctions for it are severe, then rule-guided behavior generally pays off for rational agents in each instance for which the rules prescribe behavior. Hobbes's solution to the problem of compliance is captured by his third law of nature.

Hobbes, Collective Action, and Climate Change

Although Hobbes's social contract theory does not explicitly address problems of collective action that have global dimensions, Hobbes's argument can be applied to such problems, including the problem of climate change.[6] Just as Hobbes argues with regard to agents in the state of nature, in our interdependent global world it seems rational for societies to be self-interested concerning questions of climate change, because some of the most significant human causes of climate change, in particular the emission of greenhouse gases, are strongly connected with the level of economic activity of society. *Ceteris paribus*, the greater the economic activity of a society, the wealthier the society tends to be. If a society aims to grow its economy steadily, then the society has an incentive to increase its productive capacity and, *ceteris paribus*, such increased production leads generally to an increase in emissions that negatively affect the human environment. In addition, the implementation of environmentally friendly means of production typically increases production costs and, if other societies do not implement such means and no institutional incentives, such as carbon credits, are provided that offer economic incentives to pollute less, then, *ceteris paribus*, societies that do implement such means weaken their comparative economic power.[7] In our interdependent world, societies compete with each other economically at least to some extent, and the use of environmentally friendly and typically more expensive means of production, if not enforced globally, comes with a cost that usually is not matched by a direct economic benefit.[8]

Moreover, environmental goods, such as clean air, face a common goods problem because they are non-excludable. Because the economic benefits of employing less environmentally friendly means of production are immediate and the costs of pollution are shared by many (if not all), societies do not have a direct incentive to restrict their pollution. Instead, they have an incentive to overuse the "commons" (Hardin 1968). In Hobbesian terms, once environmental regulations are in place, societies have an incentive to free-ride because it is generally difficult, and often impossible, to trace pollution precisely to the originator and demand adequate compensation for their pollution and other negative externalities that are associated with it. The tendency to free-ride is even greater in cases where the effects of pollution, or other human causes of climate change, are primarily global and lie in the future. In such cases of spatial and temporal divergence between the costs and benefits of measures to mitigate climate change, societies do not have an economic interest in enforcing environmental restrictions locally in order to prevent pollution today. Instead, they have an incentive to prioritize their national (economic) interests over the geographically dispersed and shared costs of climate change in a distant future.

In addition, as Hobbes argues with regard to agents in the state of nature, societies in our global world are, in Hobbes's weak sense, roughly equal by nature in that they are able to threaten others' wellbeing through pollution, especially through the emission of greenhouse gases. Despite the use of environmentally friendly means of production, fast-growing developing and developed societies currently pollute the environment significantly more than less developed, stagnant societies. Nevertheless, this fact does not mean that currently less developed societies do not have the ability to cause significant, permanent climate change. Considered over time, even the "weakest" societies are able to affect the environment negatively and often irreversibly so, and thus have an ability to threaten human existence in this world in the long term. Further, in practice it is often difficult to judge precisely which societies pose the biggest threat to climate change today and/or in the future, and thus, in theory all societies must be considered to be a risk with regard to climate change, although some societies may be considered to be bigger risks than others.

This consideration is significant, especially in a deeply morally pluralistic world in which substantial moral agreement on questions of climate change typically cannot be reached. As Hobbes's argument clarifies, in a deeply morally pluralistic world, agents cannot be assumed to share a common moral ground that could serve as a basis for such agreement. In a deeply morally pluralistic world, agents often reach conflicting conclusions about adequate measures to mitigate the negative effects of climate change. In fact, societies may even hold irreconcilable moral ideals, such as the moral ideals of freedom, equality, and impartiality, or irreconcilable interpretations of such ideals that inform their conceptions of distributive justice. As a result, in addition to competing economic interests, societies do not necessarily share a moral basis to address questions of climate change.

Finally, and most importantly, even if a global (minimal) consensus on measures to mitigate the negative effects of climate change through bi- or multilateral agreements could be reached (as is suggested by the Paris Agreement to which I will turn in the next section), from a Hobbesian perspective the implementation and enforcement of any such agreement are not guaranteed. In our world, *de facto* and *de jure*, no global Leviathan in Hobbes's sense exists that would have the power to enforce the demands of such an agreement globally, although some of the demands may be enforceable at the national level through the regulatory institutions of particular societies and at the global level through existing supranational institutions. Such decentralized efforts, however, typically are inefficient and often ineffectual and, without a global Leviathan, the problem of free-riding in particular prevails. In fact, according to Hobbes's argument, no binding agreements can be reached at all without a global Leviathan.

Overall, concerning the question of climate change, the central assumptions of Hobbes's social contract theory are not unrealistic for our world. The problem of climate change represents a genuine problem of collective action where individual and collective rationality conflict. As Olson (1965, 2) puts it in his seminal work on collective action:

> If the members of a large group rationally seek to maximize their personal welfare, they will *not* act to advance their common or group objectives unless there is coercion to force them to do so, or unless some separate incentive, distinct from the achievement of the common or group interest, is offered to the members of the group individually on the condition that they help bear the costs or burdens involved in the achievement of the group objectives.

In such a world, despite the fact that currently no global Leviathan in Hobbes's sense exists in our world, Hobbesian moral contractarianism seems to be an adequate moral approach to address questions of climate change, in particular if one considers the far-reaching implications that climate change bears for human existence.

Methodologically, Hobbesian moral contractarianism is also sufficiently pluralistic to accommodate both ethical and economic considerations. Some of the recent debates on climate change center on the question of whether climate change is primarily a moral or an economic problem (Gardiner and Weisbach 2016). Hobbesian moral contractarianism can reconcile these two viewpoints within one normative framework. In addition, Hobbesian moral contractarianism can accommodate feasibility concerns. Economists often criticize philosophers for offering utopian solutions that cannot be implemented in practice, whereas philosophers tend to criticize economists for being concerned only with that which is feasible, including politically feasible, even if that which is feasible is not ethically defensible. Hobbesian moral contractarianism addresses these concerns because the approach is empirically informed and ethically and politically realistic.

In the following, I clarify these considerations further, in particular the consideration that, despite the current lack of existence in our world of a global Leviathan in Hobbes's sense, Hobbesian moral contractarianism provides essential conceptual tools to address the problem of climate change in our current world. I suggest a Hobbesian solution to climate change that relies on the conceptual distinction between an "overlapping consensus" and a "*modus vivendi*."

Overlapping Consensus versus *Modus Vivendi*: From Kyoto to Paris

In the literature about the social contract, in particular the discussion of the difference between the positions of moral contractualism (Rawls 1971) and moral contractarianism (Hobbes 1651), a distinction is often made between two types of agreement on normative standards: overlapping consensus and *modus vivendi*.[9] According to Rawls (1987, 11), an overlapping consensus:

> Is affirmed on moral grounds, that is, it includes conceptions of society and of citizens as persons, as well as principles of justice, and an account of cooperative virtues through which those principles are embodied in human character and expressed in public life.

As such, an overlapping consensus is likely to ensure stable, peaceful long-term cooperation. A *modus vivendi*, by contrast, is based on the convergence of individual or group interests that typically reflect agents' bargaining power that may shift over time. According to Rawls (1993, 146–149), a *modus vivendi* is only a temporary arrangement among agents to which they reluctantly agree based on self-interested reasons in order to go on with their lives in the light of moral disagreement.

Rawls is correct to point out that an overlapping consensus on normative standards may be more stable than a *modus vivendi* in that the former is assumed to be based on substantial moral grounds and is often also socially and culturally entrenched, whereas the latter is assumed to be supported merely on instrumental grounds and thus is considered by agents as a mere means to some other end that they value. In a deeply morally pluralistic world, however, not all agents share agreement on substantial moral grounds, and thus, *de facto*, an overlapping consensus in Rawls's sense often does not exist. In this case, an agreement on the grounds of pure instrumental rationality may be the only feasible alternative to ensure peaceful long-term cooperation. Gray (2000, 139) describes the two underlying paradigms as follows:

> In political philosophy there are few timeless verities. Both Rawlsian and neo-Hobbesian liberalism are responses to modern pluralism. Rawlsian liberalism seeks to transcend pluralism by developing an agreed conception of justice. In so doing it reposes extravagant hopes in the overlapping consensus which it imagines it has found in some late modern societies. As a consequence, its real aim is the restoration of a non-existent or vanishing ethical monoculture. By contrast, the neo-Hobbesian philosophy of *modus vivendi* is well suited to societies, now and in the future, that contain many ways of life.

The finding that a *modus vivendi* may be the only feasible form of agreement on normative standards in a deeply morally pluralistic world is not necessarily problematic because, despite the fact that a *modus vivendi* is assumed to be based merely on a convergence of individual or group interests, a *modus vivendi* can be stable if (i) agents understand the purpose of the agreement, (ii) the agreement entails strict conditions under which revision of its terms is possible, (iii) the terms of the agreement are publicly known and accepted by all, and (iv) agents have an interest in enforcing the terms of the agreement institutionally in order to ensure its long-term implementation.[10] In the following, I argue that the distinction between an overlapping consensus and a *modus vivendi* is not merely philosophically interesting but also practically relevant. In fact, the distinction could be considered to underlie the change in strategy with regard to climate change negotiations from the Kyoto Protocol to the Paris Agreement.[11]

The ultimate goal of the United Nations Framework Convention on Climate Change (UNFCCC, 1992, Article 2) is to stabilize the "greenhouse gas concentration in the atmosphere at a level that would prevent dangerous anthropogenic interference with the climate system." Since the formulation of the UNFCCC as a basis for global climate change action, several attempts have been made to determine more precisely (i) the level of greenhouse gas emissions that would count as "dangerous anthropogenic interference" with our climate system and (ii) contributions from greenhouse gas emitters that would be necessary to prevent reaching such a level. With regard to the latter, from the start the negotiations were based on a particular notion of distributive justice that demands consensus among the parties to the agreement and that determines the contributions of the parties to the agreement according to the "principle of common but differentiated responsibilities and respective capabilities" (UNFCCC, 1992, 3.1).[12]

The principle of common but differentiated responsibilities and respective capabilities demands that societies reduce their greenhouse gas emissions according to their contributions that have caused and continue to cause the increase of such emissions in the Earth's atmosphere and their capacity to remedy this problem. Because a society's contribution towards this problem and its capacity to remedy it are not always the same, in particular in the case of developing countries that often face other acute challenges such as severe poverty, a bifurcated system was introduced that differentiates developed and developing countries, as is manifested in the Kyoto Protocol by the distinction between so-called "Annex-I" and "non-Annex-I" countries. In the name of justice, including notions of corrective justice, the Kyoto Protocol exempts large portions of the developing world from obligations to reduce greenhouse gas emissions and shifts the burden to do so mostly onto the developed world that, due to past emissions, is assumed to be primarily responsible for the threat of climate change.[13]

The Kyoto Protocol (1997) can be considered a milestone in global climate change negotiations and a significant step towards reaching the goal of the UNFCCC. Although full implementation of the Kyoto Protocol would not prevent all negative interference, the protocol aims to keep the increase in global temperature to less than 2°C above its pre-industrial level, which is essential to prevent dangerous anthropogenic interference with the Earth's climate system. Despite such progress, the Kyoto Protocol faces three essential problems. First, some of the most significant greenhouse gas emitters, such as the United States and also China initially, did not sign the agreement. Second, the Kyoto Protocol did not adequately consider the rapid growth of some developing societies that is accompanied by a significant increase in their greenhouse gas emissions. Third, and most importantly, the Kyoto Protocol employs a strict top-down regulation process for both the determination of the parties' contributions towards the reduction of greenhouse gas emissions and the review and compliance process.

In other words, the Kyoto Protocol assumes an overlapping consensus on questions of climate change. *De facto*, however, such a consensus does not exist, and thus, the Kyoto Protocol has led to resistance among some of its participants. Ultimately, it may even have hindered the mitigation process, or at least diverted the discussion away from determining the precise contributions of the parties to the agreement towards the mitigation of the negative effects of climate change to more foundational questions of distributive justice (Light 2016, 488–490). In addition, although defensible in theory, in particular from the perspective of Hobbesian moral contractarianism, the strictly top-down regulation process that is employed by the Kyoto Protocol, especially for the determination of the parties' contributions towards the reduction of greenhouse gas emissions, may have inadvertently discouraged voluntary participation and non-governmental support in the collective effort to reduce human causes of climate change.

The Paris Agreement that was negotiated by 196 countries in 2015 and that will be in effect in 2020 corrects these problems of the Kyoto Protocol. Despite the announced intention of the United States government to withdraw from the Paris Agreement in June 2017 and its official withdrawal from the Paris Agreement in August 2017, the agreement brings together almost all existing societies in their effort to mitigate the negative effects of climate change. In this sense, the agreement has almost universal reach and, apart from the United States, includes the most significant greenhouse gas emitters. In addition, the Paris Agreement (Article 2, §1a) sets a more ambitious goal than the Kyoto Protocol in its attempt to hold "the increase in the global average temperature to well below 2°C above pre-industrial levels and [to pursue] efforts to limit the temperature increase to 1.5°C above pre-industrial levels." Also, despite relying on the principle of common but differentiated responsibilities and respective capabilities (as does the Kyoto Protocol), the Paris Agreement has abandoned the bifurcation between developing and developed societies. Although the Paris Agreement recognizes the need to support developing countries in their effort to implement the terms of the agreement (see especially Articles 3, 4, and 10), it requires that all parties submit emission reduction plans.

Most importantly, however, the Paris Agreement does not rely on a strict top-down regulation process for both the determination of the parties' contributions and the reporting and review process. Instead, with regard to the former the agreement relies on a bottom-up approach, and with regard to the latter it relies on a top-down approach. That is, instead of internationally determined contributions, the Paris Agreement asks the parties to the agreement to state their best efforts concerning the reduction of greenhouse gas emissions in the form of "intended nationally determined contributions" and to strengthen these efforts over time. In collaborative spirit, the parties to the agreement were even asked to announce publicly their intended contributions nine months prior to the meeting in Paris in order to discuss, and potentially revise, their targets before they were recorded officially.

In philosophical terms, in contrast to the Kyoto Protocol, the Paris Agreement does not assume an overlapping consensus on questions of climate change that requires substantial moral agreement on questions of climate change among the parties to the agreement, although some parties to the agreement may be motivated by genuine moral considerations in their efforts to mitigate the negative effects of climate change. Instead, the Paris Agreement allows the parties to the agreement themselves to determine their intended contributions towards the mitigation of the negative effects of climate change based on their particular sense of distributive justice and considerations of political feasibility, and not as a result of the assumption of an overarching moral consensus on questions of climate change among the parties to the agreement. The Paris Agreement does not try to impose a global Leviathan with full legislative, executive, and judicial powers on the parties to the agreement. Instead, it allows the parties to the agreement to retain their legislative power so that they can tailor their contributions towards the mitigation of the negative effects of climate change and their implementation according to their circumstances and their specific understanding of fairness with regard to questions of climate change.

Also, in line with this consideration, the Paris Agreement relies on a holistic approach that focuses its efforts not only on mitigation, but also on processes of adaptation and support that are essential for sustainable development. The Paris Agreement (UNFCCC 2015, Article 6, §8) recognizes:

> The importance of integrated … and balanced non-market approaches being available to Parties to assist in the implementation of their nationally determined contributions, in the context of sustainable development and poverty eradication, in a coordinated and effective manner, including through, inter alia, mitigation, adaptation, finance, technology transfer and capacity building, as appropriate.

In Hobbesian terms, the Paris Agreement demands that the parties to the agreement outsource merely the reporting and review process of their nationally determined contributions. This process involves expert reviews to ensure that the progress towards fulfilling the nationally determined goals is monitored realistically and transparently. The Paris Agreement (Article 13, §1) also stipulates the creation of a new transparency system "[i]n order to build mutual trust and confidence and to promote effective implementation, an enhanced transparency framework for action and support, with built-in flexibility which takes into account Parties' different capacities and builds upon collective experience." More specifically, reporting and review processes are incorporated at all levels of the Paris Agreement and are combined with regular cycles of taking stock at the global level and renewing pledges every five years in order to allow measurement of the progress towards reaching the agreed climate change goals.

The specific architecture of the Paris Agreement has led to an interesting dynamic. Although the nationally determined contributions of the parties to the Paris Agreement are not legally binding, they are politically significant because they serve as important signals to other societies. The Paris Agreement increases the political (peer) pressure on governments and fosters ambition by allowing the parties to the agreement to increase their contributions at any time (Article 4, §11). As such, the withdrawal of a party to the agreement, as in the case of the United States, does not necessarily threaten the overall goal of the agreement. Instead, it allows other parties, such as China, the European Union, and India, to step in and strengthen their commitments, as happened at the G20 Summit in Hamburg in the summer of 2017.

In their final statement of the G20 Summit in Hamburg, the other nineteen leading nations of the world acknowledged the announced intention of the United States to withdraw from the Paris Agreement in June 2017, but they made it clear that they remain strongly committed to the agreement. Also, to mention just one example, France announced its intention to phase out the use of coal for the production of electricity after 2022 and its intention to prohibit all petrol and diesel vehicles by 2040 as part of their national goals in the context of the Paris Agreement. Such continued and increased commitment will help these societies to build their international reputations and leadership status in the world, in particular because questions of climate change have developed into a central topic in world politics. As Light (2016, 496) puts it: "What is emerging here is arguably a dynamic race to the top for climate leadership between the world's … superpowers."

The Paris Agreement links questions of climate change to questions of political leadership more generally. Returning to my quote of Olson (1965, 2), the Paris Agreement introduces a "separate incentive, distinct from the achievement of the common or group interest" to achieve its goals. The Paris Agreement introduces a political incentive to participate in climate change negotiations and, in doing so, embeds climate ethics diplomacy into broader questions of national leadership that are decisive for the long-term wellbeing of societies in our interdependent global world. In less romantic Hobbesian terms, due to the lack of an overlapping consensus on questions of climate change, the Paris Agreement instrumentalizes environmental concerns partly as a means to other ends, which, if these ends are considered to be significant, allows environmental problems to be addressed, even if some or all parties to the agreement may ultimately agree to pursue environmental goals for entirely (strategic) nonmoral reasons. In addition, this feature of the Paris Agreement allows societies to exert pressure on other societies to live up to expectations not only via "shame and blame," but also via the threat of political repercussions. In short, the parties to the Paris Agreement may decide to punish potential defectors by not cooperating with them on other important issues, in particular, economic matters, although the Paris Agreement lacks a binding legal mechanism that allows the parties to the agreement to punish other societies that do not fulfill their agreed obligations. The Paris Agreement does not merely stress the global interconnectedness of our world, but it strategically employs it to achieve its goals.[14]

Philosophically, the shift from the assumption of an overlapping consensus on questions of climate change that can be seen to underlie the Kyoto Protocol to the Hobbesian-inspired *modus vivendi* that can be seen to underlie the Paris Agreement may, from an environmental perspective, lead to the best overall outcome in the real world, even if this outcome may ultimately still fail to prevent dangerous anthropogenic interference with our climate. In terms of the Paris Agreement, the suggested Hobbesian strategy will be successful only if the combined reduction in greenhouse gas emissions is sufficient to reach the 1.5°C goal. However, because the nationally determined contributions of the parties to the agreement, as well as the reporting and review process of these goals, are transparent and public, the parties to the Paris Agreement have an incentive to determine their contributions so that together they ensure reaching the core goal of the agreement, even if they do not reach this goal in the first round of pledges. Otherwise, the individual contributions of the parties, no matter how great they may be, will be ineffective. The Paris Agreement recognizes the structure of the collective action problem that underlies climate change and offers a genuinely collective solution to this problem that provides all parties to the agreement with incentives to participate. The agreement helps to build trust and increase cooperation among its participants, which is essential to address such a large-scale collective action problem.

This collaborative spirit of the Paris Agreement is strengthened further by the fact that the agreement does not stop its involvement at the national government level. Instead, it includes local participation, in particular by regional and local officials, such as governors, mayors, and other state leaders (see in particular Article 20 of the Paris Agreement). This feature of the Paris Agreement is significant because commitments at the national government level typically require support at the state and community levels to ensure the successful implementation of such commitments. As Lashof (2015) points out:

> More than 80 jurisdictions, representing a combined GDP larger than that of the United States, have now committed to reduce their emissions by at least 80 percent or to less than 2 tons per capita by 2030. Meanwhile former New York City Mayor Michael Bloomberg convened the C40 Cities for Climate Action meeting in Paris, which now represents more than 80 cities and 25 percent of global GDP.

In addition, citizens and businesses as well as environmental, human rights, and labor organizations have become significantly involved in the Paris Agreement by committing to the reduction of greenhouse gas emissions for which they are responsible.

This grassroots backing of the Paris Agreement, which reaches down to the level where the political will is formed, increases pressure on governments to consider the actual interests of their citizens concerning climate change. Further, it raises the standards for accountability and transparency of government actions, which may help to break established patterns that favor the status quo. This aspect of the Paris Agreement is especially important because the economic impacts of climate change responses are complex and often may be seen to conflict with national interests. In order to reach the ambitious goals of the Paris Agreement, which require substantial decarbonization in the global economy and potentially even the removal of existing carbon dioxide from the atmosphere, income and wealth are likely to be redistributed because new technologies will have to be developed and implemented. This process may provide some societies with an economic advantage, in particular societies that are significantly invested in renewable energy markets and/or possess an abundance of and easy access to natural resources, such as sunlight and wind, for renewable energy.[15]

Because the economic impacts of the Paris Agreement are unclear, governments and lobbyists may have incentives to preserve the status quo that brought them into power. The Paris Agreement directly engages civil society that may have more nuanced economic and non-economic interests that are less determined by national concerns than government interests and that may allow societies to overcome some of the conservative tendencies that currently slow down the process of change.[16] In the case of the United States, for example, states, cities, and the private sector will most likely bear a significant part of the efforts to reduce greenhouse gas emissions now that the federal government has withdrawn from the Paris Agreement. Overall, the Paris Agreement includes engagement at all levels and, in doing so, adds another democratic element to global climate change negotiations. The Paris Agreement is not only more democratic than previous climate change agreements in that it allows the parties to the agreement to determine their contributions towards the mitigation of the negative effects of climate change, but it also strengthens democratic involvement at all levels.

Concluding Remarks

Climate change is one of the most pressing environmental issues of our time and, if our concerted efforts fail to address this collective action problem adequately, then our past and current actions will have significant negative effects not only on us today, but also on future generations. If the problem of climate change is approached from a realistic perspective with the goal to determine empirically sound solutions that are ethically defensible, then Hobbesian moral contractarianism is an adequate moral framework to guide our considerations, in particular in a world in which no overlapping consensus on questions of climate change exists, as discussion of the Kyoto Protocol suggests. Currently, however, the institutional structures, in particular a global Leviathan in Hobbes's sense, that would be necessary to address the problem of global climate change according to Hobbes's social contract theory do not exist in our world. Moreover, considering the history of human cooperation, or lack thereof, and the complex political landscape of our current world, we may also not want to establish such an absolute superpower.

Despite this consideration, I have argued that a Hobbesian solution in the form of a *modus vivendi* may be sufficient to address the problem of collective action that underlies climate change concerns. A Hobbesian-inspired *modus vivendi* does not assume a moral consensus on questions of climate change. Instead, it assumes merely a common end that all agents aim to reach independent of their particular goals and that may be embraced for (self-interested) economic reasons, political reasons, genuine moral reasons, or any other reasons. Economic reasons, at least in the short term, often speak against the need to mitigate the negative effects of climate change, whereas political reasons, in particular in the current political climate, tend to speak in favor of it. The Paris Agreement seems to be based on such a Hobbesian-inspired *modus vivendi* that provides solutions to both of the collective action problems that Hobbes describes, namely, the assurance problem and the compliance problem. The bottom-up approach of the Paris Agreement that relies not only on government action but also involves civil society encourages agents, step by step, to learn to trust each other with regard to their contributions towards the mitigation of the negative effects of climate change, even in the absence of a global Leviathan. The top-down review and reporting process solves the compliance problem because the parties to the agreement are bound to the goals that they set themselves. In this sense, the Paris Agreement could be considered to be guided by reasoning that underlies Hobbesian moral contractarianism.

If the overall reasons, moral or otherwise, of the parties to the Paris Agreement together with adequate peer pressure (including the possibility of excluding defectors from cooperation in general, as suggested by Hobbes) outweigh the economic costs, then all parties to the Paris Agreement have reasons to keep the agreement, even if the position of moral contractarianism shows that the free-rider problem remains acute because reducing greenhouse gas emissions is costly and the switch to renewable sources of energy may shift economic power. The hope is that such a system of trust as well as of checks and balances will be sufficient to realize the shared goal among agents. In order not to risk our future, it seems wise to start with (or return to) such realistic Hobbesian assumptions about the problem of climate change and build from there. Hobbesian moral contractarianism represents a realistic approach to climate change that does not assume that the world and its inhabitants, morally speaking, are better than they really are and coerce agents into an agreement that they do not fully embrace. Instead, Hobbesian moral contractarianism aims to implement institutional structures that ensure cooperation even under unfavorable conditions. Nevertheless, over time, when societies collaborate with each other in their efforts to mitigate the negative effects of climate change and they learn to trust each other, then they may be able to reach a genuine overlapping consensus on questions of climate change based on substantial moral grounds. If such an agreement can be reached on more demanding goals concerning the mitigation of the negative effects of climate change and the agreement is morally embedded, then all the better from a moral point of view and for the sustainability of our world.

Notes

1 For discussion of the importance of reducing especially the emission of short-lived climate pollutants, see Light and Taraska (2016).
2 The *Climate Change 2014: Synthesis Report* of the Intergovernmental Panel on Climate Change (IPCC 2014, Foreword) states that "[t]he IPCC is now 95 percent certain that humans are the main cause of current global warming."
3 In the following, I draw from my previous discussion of Hobbes's moral theory, in particular in Moehler (2018, 42–50).
4 For further discussion of Hobbes's Foole, see Vanderschraaf (2010, 37–58).
5 See Hampton (1986, 80–89), for example.
6 In the literature, Hobbes's social contract theory has been applied in particular to questions of international relations, international law, and global justice. See May (2013), for instance.
7 For discussion of the morality of tradable pollution permits, see Steidlmeier (1993).
8 For further discussion of this point, see Morgenstern (2011).
9 For detailed discussion of this distinction, see Moehler (2018, 176–177), from which my discussion draws.
10 Muldoon (2016, Chapter 4) argues that the bargaining processes that underlie *modus vivendi* agreements and their revision may help to achieve stability in diverse societies.
11 The following discussion has greatly benefited from a presentation by Andrew Light to the Program in Philosophy, Politics, and Economics at Virginia Tech (April 17, 2017). For a brief history of climate ethics diplomacy, see Light (2016) and Light and Taraska (2016).
12 For a defense of the broader view that issues of climate change should be part of a theory of intergenerational and global distributive justice, see Caney (2012).

13 For discussion of the concept of (political) responsibility in the context of climate change, see Gardiner (2011, 48–50).
14 For further discussion of possible responses to noncompliance with regard to climate change goals, see Caney (2016, 33–35).
15 For a brief history of the role that different energy sources have played in the development of human civilization, see Jamieson (2011, 16–21).
16 For an analysis of the current political inertia that prevents climate goals from being addressed swiftly and effectively, see Maltais (2016).

Bibliography

Caney, Simon. 2012. "Just Emissions." *Philosophy and Public Affairs* 40 (4): 255–300.

Caney, Simon. 2016. "Climate Change and Non-Ideal Theory: Six Ways of Responding to Non-Compliance." In *Climate Justice in a Non-Ideal World*, edited by Clare Heyward and Dominic Roser, 21–42. Oxford: Oxford University Press.

Gardiner, Stephen. 2011. "Is No One Responsible for Global Environmental Tragedy? Climate Change as a Challenge to Our Ethical Concepts." In *The Ethics of Global Climate Change*, edited by Denis Arnold, 38–59. Cambridge: Cambridge University Press.

Gardiner, Stephen, and David Weisbach. 2016. *Debating Climate Ethics*. Oxford: Oxford University Press.

Gauthier, David. 1986. *Morals by Agreement*. Oxford: Clarendon Press.

Gray, John. 2000. *Two Faces of Liberalism*. New York: New Press.

Hampton, Jean. 1986. *Hobbes and the Social Contract Tradition*. New York: Cambridge University Press.

Hardin, Garrett. 1968. "The Tragedy of the Commons." *Nature* 162 (3859): 1243–1248.

Hobbes, Thomas. 1651. *Leviathan*. Edited by Richard Tuck. 1996 ed. Cambridge: Cambridge University Press.

IPCC. 2014. *Climate Change 2014: Synthesis Report. Contribution of Working Groups I, II and III to the Fifth Assessment Report of the Intergovernmental Panel on Climate Change*. Edited by Core Writing Team, Rajendra Pachauri and Leo Meyer. Geneva, Switzerland: IPCC.

Jamieson, Dale. 2011. "Energy, Ethics, and the Transformation of Nature." In *The Ethics of Global Climate Change*, edited by Denis Arnold, 16–37. Cambridge: Cambridge University Press.

Kavka, Gregory. 1986. *Hobbesian Moral and Political Theory*. Princeton, NJ: Princeton University Press.

Lashof, Dan. 2015. "The Four Pillars of Progress in Paris." *NextGen Climate America*. https://nextgenpolicy.org/blog/the-four-pillars-of-progress-in-paris (no longer active).

Light, Andrew. 2016. "Climate Diplomacy." In *The Oxford Handbook of Environmental Ethics*, edited by Stephen Gardiner and Allen Thompson, 487–500. Oxford: Oxford University Press.

Light, Andrew, and Gwynne Taraska. 2016. "A Responsible Path: Enhancing Action on Short-Lived Climate Pollutants." In *Climate Justice in a Non-Ideal World*, edited by Clare Heyward and Dominic Roser, 169–188. Oxford: Oxford University Press.

Maltais, Aaron. 2016. "A Climate of Disorder: What to Do About the Obstacles to Effective Climate Politics." In *Climate Justice in a Non-Ideal World*, edited by Clare Heyward and Dominic Roser, 43–63. Oxford: Oxford University Press.

May, Larry. 2013. *Limiting Leviathan: Hobbes on Law and International Affairs*. Oxford: Oxford University Press.

Moehler, Michael. 2009. "Justice and Peaceful Cooperation." *Journal of Global Ethics* 5 (3): 195–214.

Moehler, Michael. 2018. *Minimal Morality: A Multilevel Social Contract Theory*. Oxford: Oxford University Press.

Moehler, Michael. 2020. *Contractarianism*. Cambridge: Cambridge University Press.

Morgenstern, Richard. 2011. "Addressing Competitiveness in US Climate Policy." In *The Ethics of Global Climate Change*, edited by Denis Arnold, 216–231. Cambridge: Cambridge University Press.

Muldoon, Ryan. 2016. *Social Contract Theory for a Diverse World: Beyond Tolerance*. New York: Routledge.

Olson, Mancur. 1965. *The Logic of Collective Action: Public Goods and the Theory of Groups*. 1971 ed. Cambridge: Harvard University Press.

Rawls, John. 1971. *A Theory of Justice*. 1999 revised ed. Cambridge: Harvard University Press.

Rawls, John. 1987. "The Idea of an Overlapping Consensus." *Oxford Journal of Legal Studies* 7 (1): 1–25.

Rawls, John. 1993. *Political Liberalism*. New York: Columbia University Press.

Steidlmeier, Paul. 1993. "The Morality of Pollution Permits." *Environmental Ethics* 15 (2): 133–150.

UNFCCC. 1992. *The United Nations Framework Convention on Climate Change*. https://unfccc.int/files/essential_background/background_publications_htmlpdf/application/pdf/conveng.pdf.

UNFCCC. 2015. *Paris Agreement*. http://unfccc.int/files/essential_background/convention/application/pdf/english_paris_agreement.pdf.

Vanderschraaf, Peter. 2010. "The Invisible Foole." *Philosophical Studies* 147 (1): 37–58.

8

NATURAL LAW THEORY AND CLIMATE CHANGE

Colleen McCluskey

Natural Law

Natural law theories are a diverse group.[1] Nevertheless, they are all committed to two fundamental ideas. First, they maintain that there is a set of intrinsically valuable goods that in turn define a flourishing human life. Second, they argue for a priority of the good over the right (see Crowe 2011, 297–301; Murphy 2001, 40; Murphy 2011, 1.2–2, 2.2–3). The set of goods specified by a particular theory forms the basis for the derivation of fundamental moral principles or rules that serve as a guide for practical rationality. All natural law theorists have these commitments in common, although they may add further commitments, as I discuss below.

The set of human goods defined by natural law theorists are thought to be fundamental and non-reducible (Crowe 2011, 298; Finnis 2011, 92–93). Together as a group, these goods are seen as both necessary and sufficient. Many theorists ground these goods in an account of human nature.[2] In addition to being intrinsically valuable, these goods are held to be constitutive of human flourishing and perfective of human nature. As such, they play important roles in human well-being. Murphy, for example, argues for a very tight connection between the intrinsic worth of the basic goods and their roles in human flourishing. Because they are grounded in human nature, the goods that are objectively worth pursuing (his description for intrinsically valuable human goods) just are those goods that define flourishing (see Murphy 2001, 3 and 96–138).

There is both disagreement and commonality among theorists over the specification of these goods. For example, Finnis includes life, knowledge, play, aesthetic experience, sociability (including friendship), practical reasonableness, and religion (Finnis 2011, 86–90). Murphy, who bases his account of the goods in part on that of Finnis (Murphy 2001, 265, note 1), lists the following as basic goods: life, knowledge, aesthetic experience, excellence in play and work, excellence in agency,[3] inner peace, friendship and community, religion, and happiness (Murphy 2001, 96). Gómez-Lobo has the following on his list: life, family, friendship, work and play, aesthetic experience, theoretical knowledge, and integrity (Gómez-Lobo 2002, 10–25). Chappell includes friendship, aesthetic value, pleasure and the avoidance of pain, physical and mental health and harmony, rationality and reasonableness, truth and knowledge, the natural world, people, fairness, and achievements (Chappell 1998, 43).[4] Some theorists argue that these lists are at least in principle revisable (see Murphy 2001, 101), while others hold that they are not and could not be exhaustive (see Chappell 1998, 44). Some concede the latter point but suggest that additional goods either would not be basic or would constitute a mode of pursuit (to use Finnis's term) for a basic good (see Finnis 2011, 90–92).

The second commitment, that of the priority of the good over the right, follows from the idea that the basic goods are foundational for action (see Crowe 2011, 300; Gómez-Lobo 2002, 41–47; Chappell 1998, 74–84). Basic goods are intrinsically valuable and as such also enhance or constitute a valuable human life. Since, on these views, all human beings desire their own flourishing, basic goods provide reasons for their pursuit or at least reasons not to violate them. Their importance generates a system of underlying principles that regulates our relationship with the basic goods (Crowe 2011, 300). These principles define what is called practical reasonableness (Crowe 2011, 300; Murphy 2001, 2–3). What is reasonable or rational for human beings to do is to incorporate the basic goods into their lives or at the very least not damage those goods. Nevertheless, different theorists have differing understandings of the requirements of practical reasonableness, which in turn can lead to divergent judgments on particular moral issues (Crowe 2011, 301).

Natural law theorists have often taken as their foundation Thomas Aquinas's remarks about natural law in his so-called treatise on law in *Summa theologiae* (see *ST* I-II.91 and 94).[5] Aquinas (1882) argued that the first principle of the natural law was to do good and avoid evil (*ST* I-II.94.2). Subsequent principles rest upon this general principle insofar as what practical reason apprehends as good is judged as suitable for pursuit while what practical reason apprehends as bad is judged as something to be avoided (*ST* I-II.94.2). The content for these principles is given by particular natural inclinations common to all human beings. On Aquinas's view, these inclinations include self-preservation, reproduction of the species, pursuit of knowledge (specifically for Aquinas, knowledge about God), and community with others (*ST* I-II.94.2). These inclinations, which are a function of our nature, determine what is good for human beings. Fulfillment of these goods constitutes our flourishing. To pursue such goods is rational for human beings, for it reflects the actual desires of individuals, the satisfaction of which human beings naturally incline toward (see *ST* I-II.94.3; *ST* I-II.18.6). Obviously much more would need to be said in terms of developing this account; for example, Aquinas is well aware that human beings could be inclined toward all sorts of immoral things.[6] Nevertheless, for Aquinas, there is some fact of the matter about what constitutes a genuinely fulfilling human life, and this understanding forms the basis for the derivation of the individual precepts of the natural law (*ST* I-II.94.2).

Murphy has argued that due to Aquinas's prominence in the development of natural law theories, paradigmatic accounts require substantial commitment to the fundamental components of Aquinas's view. This includes on Murphy's view a commitment to theism (Murphy 2011, 1–1.2).[7] He argues that paradigmatic forms of natural law theory are incompatible with atheism, deism, and agnosticism insofar as for Aquinas, the natural law is grounded in the relationship between God's providence, which is expressed in what he calls the eternal law, and human participation in God's providence, which he terms the natural law itself (Murphy 2011, 1.2). Murphy lists seven components of paradigmatic views:

> To summarize: the paradigmatic natural law view holds that (1) the natural law is given by God; (2) it is naturally authoritative over all human beings; and (3) it is naturally knowable by all human beings. Further, it holds that (4) the good is prior to the right, that (5) right action is action that responds nondefectively to the good, that (6) there are a variety of ways in which action can be defective with respect to the good, and that (7) some of these ways can be captured and formulated as general rules.
>
> *Murphy 2011, 1.4*

On Murphy's view, paradigmatic natural law theories include a commitment to all seven points. Theorists who reject some of these points may still be considered at least in the neighborhood as long as they accept most of them, although Murphy takes them to hold deviant theories (Murphy 2011, 1). As he notes, there are some non-theists who accept all but the first point; Gómez-Lobo is one example (see Gómez-Lobo 2002, 24–25).

The uniqueness of Murphy's conception of natural law theory is found in the first requirement, namely, a commitment to theism. It is not obvious to me that this is in fact a necessary condition for paradigmatic natural law theory. Although it is true of course that Aquinas's worldview is theistic, and the content of Aquinas's account rests on a commitment to the existence of God, still it is not obvious that an atheist could not adopt the basic framework of a natural law theory. An atheist would not accept the idea that there is a relationship between eternal law and natural law and that human beings participate in God's plan for creation, but on my view, the objectivity of natural law theory does not require such a commitment. An objectivist about ethics could hold that there are basic goods for human beings insofar as they are human beings, which goods help to define the limits of practical rationality from which the basic moral code is derived. None of this necessitates a commitment to theism. In fact, once Murphy begins to talk about these ideas, the notion of God drops out.

Climate Change

Currently, a literature search on what natural law theorists have to say about human-induced climate change reveals no published work. Furthermore, Crowe has argued that diverse understandings of the basic goods and the notion of practical reasonableness imply that different theorists are likely to arrive at divergent views on various issues (Crowe 2011, 300–301). Just as one can find consequentialist arguments for and against capital punishment, for example, so too there is no guarantee that there will be unanimity among natural law theorists over the issue of climate change.

Nevertheless, it is open to all natural law theorists to approach the topic from the standpoint of a potential threat against one or more basic goods. If climate change adversely affects a basic good or threatens human well-being enhanced by that good, human beings have a prima facie reason to do something about it. In this chapter, I discuss three ways in which climate change can constitute a danger to basic goods: from the perspective of what I will call human goods; the natural environment; and the common good. First, as I discussed earlier, for natural law theorists, basic goods have a direct connection to human nature and/or flourishing and therefore are often called human goods. I will examine the impact of climate change on individual human goods related to human flourishing. Second, one theorist includes the natural environment as a basic good, namely, Chappell (1998, 43). It follows from his view that the threat posed by climate change to the natural world requires a response. Finally, I will consider the effects of climate change on the common good. Among the natural law theorists I survey for this chapter, only Murphy includes the common good (as friendship and community good) on his list.[8] I discuss each of these three possible approaches, beginning with the threat to explicitly human goods.

Threats to Basic Human Goods

Climate change jeopardizes a number of basic human goods listed by natural law theorists. Perhaps the most fundamental is life itself, which is included in some form on everyone's list. Life as a basic good is taken to refer to human life. For some theorists, it includes not only the possession and protection of one's life qua life but also procreation (Finnis 2011, 86) or the maintenance of health (Murphy 2001, 101; Gómez-Lobo 2002, 12; Chappell 1998, 43). Climate change threatens human life and health in a number of ways. First, I shall say a few words about the underlying mechanism of climate change and then discuss its effects on the quality of human life.

Climate change is linked to an increase in greenhouse gases, the three most important of which are carbon dioxide, methane, and nitrous oxide (Berners-Lee et al. 2012, 184; Houghton 2009, 35). These gases occur naturally and help to maintain a hospitable climate for the living entities that inhabit Earth (Gardiner 2010,

4; Houghton 2009, 19–22; Henson 2014, 27–30). Many human activities contribute to the production of these gases, especially the burning of fossil fuels, but also deforestation, various agricultural practices, and the management of landfills (Berners-Lee et al. 2012, 184; Karl et al. 2009, 14; Gardiner 2010, 5–6; Houghton 2009, 204). The increasing production of these gases by human activities is the focus of concern, since their accumulative levels promote a greenhouse effect and push ambient temperatures upward (see Houghton 2009, 70–78). Carbon dioxide is especially worrisome since each molecule produced typically has a hundred-year-plus lifespan (Henson 2014, 31). Production of these gases also increases the amount of water vapor in the air, which contributes to climate change (Karl et al. 2009, 15). Increasing levels of greenhouse gases stimulate increases in air surface temperatures, which in turn have a number of complicated effects, many of them detrimental to human life (Karl et al. 2009, 17–18). On the other hand, as I illustrate below, not all of the effects of climate change threaten human life. The situation is quite complex, which affects what natural law theorists will have to say.

The most direct result of higher temperatures is an increasing trend toward heat waves. Heat waves are often discussed in terms of their toll on human lives (Houghton 2009, 214; McMichael et al. 2006, 861–62). For example, in the summer of 2003, much of Europe sweltered in a heat wave to which somewhere between 35,000 and 50,000 deaths were attributed (Henson 2014, 59; Houghton 2009, 215). Russia was hit with prolonged summer heat in 2010, also killing thousands (Henson 2014, 59, 65). As temperatures increase, it is projected that the number of heat waves will increase (Henson 2014, 72–73; Houghton 2009, 155). On the other hand, warmer winters will likely result in fewer deaths (Houghton 2009, 214; McMichael et al. 2006, 860 and 864). Rising temperatures have variable effects on human life and welfare.

Pollution produced by sunlight and stagnant conditions present during a heat wave is an additional risk to human beings. Ozone and tiny particulate matter generated from heavy metals, sulfates, nitrates, and other materials are both lung irritants and are produced in such environments. Both contributed significantly to the death tolls in the European heat wave, especially for individuals with pre-existing conditions; 20–40% of the deaths were attributed to increased ozone and particulate matter (Henson 2014, 63). Other forms of pollution, such as soot, contribute to increasing temperatures (Karl et al. 2009, 14). One study found that soot also increases Arctic ice melt (Henson 2014, 102, 118) as well as glacial melt in other parts of the world (Henson 2014, 132). Soot is produced from the combustion of biomass and fossil fuels.

Climate change is implicated in increasing both floods and droughts in various parts of the world because of its influence on rainfall, although other factors are involved as well (Henson 2014, 80–86, 94; Houghton 2009, 155–60). Hurricanes (discussed in more detail below) produce heavy rains, which often cause flooding. Deforestation and development can exacerbate flooding (Henson 2014, 86) while fluctuations in sea surface temperatures due especially to the El Niño and La Niña phenomena can play roles in the occurrence of both floods and drought (Henson 2014, 90–95). Flooding threatens human life directly and damages property and agriculture. Persistent drought affects water supplies and crops. Although material possessions are not basic goods for natural law theorists, a baseline of material assets enables human beings to pursue other basic goods such as knowledge and aesthetic experience, which are found on everyone's list. When one must work toward rebuilding one's house or a community must work toward rebuilding the infrastructure damaged by flood, for example, people lack the time, resources, and energy to devote to these other pursuits.

Melting Arctic ice has a detrimental effect both on human and on non-human life, in particular, polar bears, seals, and birds (which I discuss later in this chapter). Arctic ice provides a buffer for indigenous coastal communities, protecting them from erosion and flooding caused by winter storms (Henson 2014, 108–109). These communities have been located on the coasts for centuries, using the sea ice for access to whales and seals, upon which they depend for sustenance (Henson 2014, 105, 108). One possible solution for protecting these communities especially from the threat of winter storms would be to move them inland. Moving these well-established communities, however, threatens their culture and way of life, which if not a basic good contributes to the attainment of other basic goods such as sociability (on Finnis's list) and friendship and community (on everyone's list). Furthermore, such a move would impair the ability of these communities to sustain themselves and would be very expensive (Henson 2014, 108–109).

Increasing temperatures are harming glaciers outside of the Arctic as well. Glaciers are receding on virtually every continent, from the tropics (e.g., Mount Kilimanjaro and the Andes Mountains) to the northern hemisphere and central Asia (Henson 2014, 125–40). This in turn can affect the prospects of those who depend upon them as their water source and for hydro-electric power, such as the residents of Lima, Peru (Henson 2014, 130; Houghton 2009, 192) and the region of the Himalayas in India (Henson 2014, 132–33; Houghton 2009, 192). Water is of course essential for human life. While Lima and other communities that depend upon hydro-electric power could adapt and develop other sources of power, this in turn would require monetary and other resources, which could be burdensome for these communities, interfering with the pursuit of basic goods.

The level of the oceans has been rising with the concurrent rise in temperature. In part, their rise can be traced to the Arctic melt-off (Henson 2014, 145; Houghton 2009, 78, 176). The increase in ocean temperature also contributes to a rise in sea level (Henson 2014, 141; Houghton 2009, 176). Rising ocean levels threaten inhabitants of coastal regions from Bangladesh to the Florida Keys (Henson 2014, 153) and are well on their way to destroying island nations such as Tuvalu and the Maldives (Henson 2014, 148–51).

The development of hurricanes is highly complex and depends upon factors other than sea temperature, making it difficult to gauge the effects of climate change upon their formation (Henson 2014, 178, 180–81). Some researchers have found that while there may be the same number or fewer hurricanes/cyclones, those that do form are likely to be more powerful (Henson 2014, 182–83). At least one researcher argues that continued climate change increases the chances that more storms will intensify shortly before making landfall during the course of this century (Emanuel 2017, 500). This in turn will likely cause harm to human populations living in coastal areas, depending upon where such storms make landfall.

It is expected that the distribution of insects will be affected by climate change, including vectors for human diseases (Henson 2014, 209; Houghton 2009, 214). For example, the mosquitos that carry malaria might shift their ranges; areas where malaria was previously endemic might become inhospitable to the mosquitos, which then are apt to migrate to previously malaria-free areas. Malaria is already a major health risk globally; climate change patterns threaten to increase its range and potential damage to human health (Henson 2014, 212; Houghton 2009, 214; McMichael et al. 2006, 862). Outbreaks of West Nile virus are correlated with increasing summer heat (Henson 2014, 213). Insects may also affect plant life, including common human foods, such as the European potato crop (Henson 2014, 213). This in turn could impair the ability of local communities to sustain themselves.

Although some evidence indicates that increased levels of carbon dioxide might boost plant growth, the effect appears to be limited and not accessible to all plants (Henson 2014, 219–21; Houghton 2009, 199). Furthermore, some of the plants that appear to benefit the most are not the most beneficial to human communities; these plants include kudzu, which is considered invasive in North America, and poison ivy, which causes intense misery to those who are susceptible to it (Henson 2014, 222).

Climate change is expected to have a negative impact on agriculture in Africa and parts of Asia, Europe, and North America. As warmer temperatures result in previously colder areas becoming more suitable for farming, the northern parts of Asia, Europe, and North America may experience an agricultural boom while the southern areas suffer from heat and drought (Henson 2014, 225; Houghton 2009, 199–202). Warmer temperatures are expected to produce longer growing seasons and potentially larger areas amenable to particular crops such as rice or wheat in the developed countries of the northern hemisphere (Henson 2014, 223; Houghton 2009, 202). On the other hand, climate-induced drought and periods of extreme heat in the Global South are expected to contribute to a decline in agricultural productivity (Henson 2014, 223, 225; Houghton 2009, 199–202). This in turn will adversely affect not only the available food supply in the Global South but also those whose livelihoods depend upon agriculture. Africa is thought to be especially vulnerable, although some of this vulnerability is due to pressures other than climate change (Houghton 2009, 216).

While evidence indicates that the adverse effects of climate change on northern agriculture are likely not to be as severe as in the south, still the northern hemisphere is not entirely exempt. The 2010 heat wave across Russia contributed to the failure of Russia's wheat crop that year (Henson 2014, 65). A premature warm-up in March 2012 set up crop failures across the central section of the United States when the blossoms on fruit trees died in the subsequent April frosts (Henson 2014, 66). Southeastern Canada has seen a boost in maple syrup production while New England, traditionally the heart of the maple syrup industry, has experienced a noticeable decline (Henson 2014, 218). Although maple syrup is not required for human sustenance, still this development affects the livelihoods of those working in the New England maple syrup industry. This in turn challenges such basic goods as excellence in work (on both Murphy's and Gómez-Lobo's lists) and Chappell's good of achievement, which includes running a successful business (Chappell 1998, 41).

Climate change is expected to have an impact on the supply of fresh water. Water of course is essential for human life and for many of the activities that support human life (e.g., agriculture, industry, personal care; see Houghton 2009, 188). Climate change alters precipitation patterns, which in turn affect the amount of water available for human needs; some areas will see increases while others will experience decreases (Karl et al. 2009, 41–42, 47; Houghton 2009, 190–94). Additional factors also contribute to both the quality and the quantity of available water, including population increases, pollution, inefficiencies in irrigation in agriculture, deforestation, and an aging delivery infrastructure (Karl et al. 2009, 47–49; Houghton 2009, 194). These factors are not necessarily related to climate change, but they can exacerbate an already tenuous situation. It is expected that people's access to water will increase in vulnerability with the intensification of climate change (Karl et al. 2009, 49; Houghton 2009, 192–94).

Thus, climate change has negative effects on human goods in a number of ways. Individual climate events, such as heat waves, droughts, severe hurricanes, flooding, and changes in insect ranges can lead directly to human mortality. Other effects will be indirect. Harm to agriculture could decrease the ability of local populations to nourish themselves. People's livelihoods are likely to be disrupted, especially for those who work in agriculture (see Houghton 2009, 202); Murphy, Chappell, and Gómez-Lobo all consider work to be a basic good. A disruption in the ability of individuals to earn a living, if severe enough, can threaten their lives as well. It can also interfere with their ability to achieve other basic goods such as knowledge acquisition and aesthetic experience. Individual climate events such as hurricanes or flooding cause disruptions to peoples' lives. These disruptions in turn challenge their access to inner peace (on Murphy's list) and mental health and harmony (on Chappell's list). Thus, climate change challenges the pursuit of basic goods related to human life in a number of ways. This, in turn, gives people reasons to address and alleviate the root causes of human-induced climate change on a natural law account of ethics.

On the other hand, many of the activities that produce the greenhouse gases that fuel climate change contribute to or are essential for human survival, including agriculture, forestry management, and home heating/cooling. Industry provides individuals with the means of supporting their lives. Personal transportation enables individuals to get to their jobs and pick up supplies that maintain their lives. Even if we eliminated consumer inessentials, it is not possible to eliminate all of the activities that fuel climate change.[9] Furthermore, such structural change will at least in the short run disrupt the ability of individuals to lead flourishing lives. Thus, there is a trade-off in terms of human goods: on the one hand, human activities fuel climate change, which negatively affects human well-being, while on the other hand, these same activities are necessary for human well-being.[10]

A further complication is that not all of the news about climate change is negative. As I mentioned earlier, with warmer winter temperatures, deaths due to extreme cold are likely to decrease. The extent to which climate change threatens human well-being can also be reduced by adaptive strategies. For example, farmers in Peru have been modifying the crops they grow depending upon anticipated yearly variations in rainfall (Houghton 2009, 198). Changes in irrigation practices have been implemented in response to decreased rainfall and increased demand (see Houghton 2009, 195). The success of such strategies gives natural law theorists less reason to argue for the direct amelioration of climate change. There is no doubt that the challenges will be formidable. Given that a significant number of populations will find it difficult to produce enough food to feed themselves, major food redistribution programs would have to be put in place. Coastal residents will in all likelihood need to be relocated. These responses involve tremendous costs, both monetary and non-monetary, and necessitate serious planning and implementation. Natural law theorists, however, are not consequentialists; they are not required to add up the costs and benefits of these various approaches. What matters to them is the threat to basic goods; natural law theory requires us either to pursue these goods or at the very least not to harm these goods. To the extent that adaptive strategies are possible and able to protect human goods, natural law theory is not compelled to accept the conclusion that we ought to address the causes of climate change directly instead of availing ourselves of these other strategies, despite their costs.

Nevertheless, one might argue that a more environmentally sustainable approach to human activity would better promote and safeguard human goods than simply ameliorating the effects of climate change (on this issue, see Houghton 2009, 240–51). Once again, there are a number of complications. For example, much work in environmental ethics seems to presuppose a tight link between protecting the environment and moderating climate change. For example, Rottman et al. (2015, 134) make a direct connection between these issues in their opening paragraph. Gifford and Comeau (2011) measured the effectiveness of framing messages on climate change, the solutions to which were framed in terms of the environment. Aufrecht (2017) makes similar connections. It is not obvious, however, that simply addressing the former will take care of the latter. Some practices can be executed in an environmentally sustainable manner that nevertheless contributes to climate change. For example, rice is often taken to be an important part of a whole-food, plant-based diet, and some have argued that such diets contribute less to climate change (see for example, Castañé and Antón 2017; Berners-Lee et al. 2012). Yet a by-product of raising rice using traditional methods is the emission of methane, one of the greenhouse gases that contribute to climate change (Oo et al. 2018, 148). It is likely that a rice-based diet that does not include meat of any sort, especially from ruminants, contributes less methane than those diets that involve meat and dairy products (for comparisons of different kinds of diets, see Bryngelsson et al. 2017; Berners-Lee et al. 2012). Furthermore, some methods of rice production produce lower amounts of methane (see Oo et al. 2018, 155–58). Still, if everyone who currently consumes products from ruminants switched to a whole-food, plant-based diet that included rice, we would not thereby eliminate methane. Rice production can be done in an environmentally sustainable manner, and yet it can have a non-negligible effect on climate change.[11] Nevertheless, living in a more sustainable manner would contribute to the direct mitigation of climate change (Houghton 2009, 270–73, 393–94). For example, living in a more sustainable manner would help to decrease pollution from industry and other human activities that exacerbate climate change (see Karl et al. 2009, 53).[12]

One might also argue for the mitigation of climate change from the fact that if we continue on our present trajectory, future generations will certainly suffer. Among the natural law theorists I have surveyed for this project, I find no significant discussion of the well-being of future generations as a good to be considered. Gómez-Lobo argues that family is a basic good, but he limits his discussion to one's present family (see Gómez-Lobo 2002, 13–14). Chappell includes people as a basic good but once again does not consider explicitly those who will exist in the future, although nothing in his theory rules out this consideration (Chappell 1998, 40). Murphy says nothing at all on this topic. Finnis mentions procreation, which he argues falls under the good of life (Finnis 2011, 40). He also discusses the good of family, which falls under the category of sociability, but his remarks have to do with cherishing and educating one's children, which does not address the present issue (see Finnis 2011, 87 and 136). Maintaining the well-being of one's children could give one a reason to address climate change for those who regard family as a basic good. The chain of concern would continue once one's children had children, but I find nothing in these theories that enables us to arrive at the conclusion that we should be so far-sighted as to care about the damage our current activities will have on the prospects of future generations.

Threats to Non-Human Goods

A second natural law approach is open to Chappell, who argues that the natural world has value in and of itself.[13] Insofar as climate change adversely affects the natural environment, this gives human beings reasons to lessen its impact on the natural world over and above the reasons they have to lessen its impact on themselves.

The effects of climate change on Arctic ice have repercussions for non-human animals. Arctic ice provides habitat and easier access to food for both polar bears and ringed seals, their primary prey (Henson 2014, 106–107). Arctic birds are also affected, since some species feed upon the carcasses of seals left behind by the polar bears (Henson 2014, 108). Other bird species will be forced to shift their ranges as the fish on which they depend move in response to changing ocean environments (Henson 2014, 108). Changing Arctic landscapes impair the ability of caribou to locate their preferred food sources (Henson 2014, 113–14).

Climate change destabilizes permafrost, which covers significant sections of northern Canada, Alaska, and Russia as well as smaller sections of Scandinavia (Henson 2014, 110). As permafrost melts, it releases water into the ground, creating a marshy landscape that undermines building foundations, produces sinkholes, and disrupts the stability of trees (Henson 2014, 110–11). The melting permafrost releases substantial amounts of carbon dioxide and methane as well, further contributing to the causes of climate change (Henson 2014, 111–12).

Warmer oceans affect the living beings that depend upon them. Some studies have demonstrated a decrease in phytoplankton populations as the oceans have warmed (Henson 2014, 165). This in turn can decrease the fish populations that depend upon phytoplankton as a source of food as well as the fish and birds that prey upon the plankton eaters (Henson 2014, 165). Furthermore, the oceans absorb significant amounts of the carbon dioxide produced by human activities; this carbon dioxide is acidifying the oceans (Henson 2014, 166). Acidification in turn harms such ocean life as oysters and pteropods (Henson 2014, 166–67; Houghton 2009, 211). Coral reefs are damaged both by warming and by acidification (Henson 2014, 166, 168; Houghton 2009, 211).

Climate change is expected to impact the survival prospects for many living things. Even small changes in temperature can threaten the viability of individual species within particular ecosystems, leaving some species especially vulnerable, resulting in losses or changes in the biodiversity of that system (Houghton 2009, 203). Amphibians and reptiles are particularly at risk because regulation of their body temperatures depends upon their ability to seek out congenial environmental conditions. Increasing temperatures may induce them to move to cooler locations (Henson 2014, 199). But such migrations are often hindered by the presence of human impacts on the landscape, such as towns, cities, cultivated fields, highways, and the like (Henson 2014, 208). Heat and drought can take their toll on monarch butterflies (Henson 2014, 201). Climate change can bring about changes in the behavior of plants and animals, threatening the viability of other creatures that depend upon them for food (Henson 2014, 202–203). For example, an earlier seasonal die-off of a plant favored by the Bay checkerspot caterpillar has led to its disappearance in northern California (Henson 2014, 203).

Trees in particular are vulnerable to climate change. They grow and mature more slowly than other plants and often are viable only within a narrow environmental range as they are sensitive to fluctuations in precipitation and temperature (Houghton 2009, 204). A greater number of forest fires contributes to increasing levels of greenhouse gases as well as decreasing the number of trees available to sequester carbon dioxide (Henson 2014, 216). During the 2003 heat wave in Europe, forest fires broke out across southern, central, and eastern Europe (Houghton 2009, 215). A warming Arctic also fuels forest fires in the north. Both drier summers and increases in insect populations stress and damage individual trees, leaving forests more vulnerable to fire (Henson 2014, 112–13). Insects that are generally kept in check by cold winters are proliferating with warmer temperatures and shorter winters. Together with drought and forest fires, they are contributing to the death of large sections of forests in North America (Henson 2014, 215–16; Houghton 2009, 206).

Thus, scientists have documented that human-induced climate change has significant negative effects on the natural world. But this in itself is not sufficient to argue on the basis of natural law theory that human beings ought to ameliorate its effects. First of all, the effects of climate change are not always bad for particular species. Climate change could turn out to be beneficial for some animals that will be able to expand their ranges. This is already happening with some species of deer in North America (Henson 2014, 208). Climate change can also set up conditions under which non-native plant species enter into a particular habitat and crowd out the native flora, a success from the standpoint of the invasive species (albeit at the expense of the non-invasive species). Kudzu is a famous example. First introduced at the 1876 Centennial Exposition in Philadelphia and at the 1883 Exposition in New Orleans, kudzu was widely used as an ornamental plant and for erosion control and animal fodder in the first half of the twentieth century until its recognition as problematic in the 1950s and 1970s (Forseth and Innis 2004, 402). It has now extended its range northward into southern Canada (Henson 2014, 209).

Second, the natural world has always undergone change. In the far distant past, the climate alternated between periods of warmth and ice ages, all of which have had a significant impact on the bio-systems of the world. These changes in climate come about by virtue of natural processes (Houghton 2009, 85–87). Additional features of the Earth bring about alterations in the natural world. For example, the crashing of tectonic plates is a major contributor to changing coastlines in parts of Alaska and Japan (see Henson 2014, 144). Other factors help to maintain balance in the environment. For example, although fire and insects can be destructive to forests, they also help to maintain healthy forests; some tree species depend upon fire in order to proliferate (Henson 2014, 216). Therefore, some have argued that it is not immediately apparent that human-induced climate change harms the basic good that is the natural world.

While it is true that human-induced climate change mimics the effects of many natural processes, it does so at an accelerated pace. In general, natural processes shape conditions on Earth more slowly, giving fauna and flora time to adjust to the changes and more likely insure their survival (see Houghton 2009, 203, 241). The current pace of climate change threatens the biodiversity of Earth (Houghton 2009, 210–11). Of course many species have become extinct over the course of Earth's existence independently of human activity, but if the natural world has value in and of itself, then producing changes that have as one of their consequences the extinction of species that otherwise would have survived or even flourished is to engage in behavior that threatens a basic good. On Chappell's account, it would follow that *ceteris paribus* such activities would be immoral.

Two other approaches align themselves with Chappell's view of the natural world's value and would strengthen the argument from a natural law perspective. Both emphasize the value of the Earth over and above its instrumental worth to human beings, and both emphasize the place that human beings occupy within the natural environment. Vandana Shiva argues that in traditional Indian cosmology, nature is animated by a force called *Prakriti* (Shiva 1989, 38). *Prakriti* is the fundamental creative and powerful force in the cosmos, responsible for the creation, renewal, and sustaining of life in all of its diversity (Shiva 1989, 38–39). Human beings have a place within this system and cannot detach themselves from it. The good of the created world and the human good are interwoven and cannot be separated; in a healthy life, human beings engage in a "living, nurturing relationship" with the created order (Shiva 1989, 39). They are inherently connected; harm done to the Earth is harm done to humanity. James Lovelock argued for a similar relationship based on the inherent dynamism between living things and their environment, which produces a complex, integrated system that Lovelock argued constitutes an organism in and of itself (see Houghton 2009, 243–44; Levine 1993, 88–89).

Both of these approaches, though controversial, suggest that nature (which includes human beings as an integral part) has value over and above its instrumental importance for human concerns.[14] This discussion suggests that natural law theorists ought to follow Chappell's lead and regard the natural world as a basic good. Natural law theorists ought to hold that human beings have the responsibility to maintain the welfare of the natural environment. Insofar as climate change threatens its well-being, natural law theorists have reason to address climate change.

Threats to the Common Good

The third approach to climate change that I discuss in this chapter involves Murphy's understanding of what he calls the good of community.[15] Murphy argues that human beings achieve the good of community when agents come together for the sake of pursuing a common end qua common end (Murphy 2001, 126). He describes a common end in this way in order to rule out cases where agents pursue the same end but for their own idiosyncratic reasons or where agents pursue the same end but in isolation from one another. For example, my neighbor and I both go grocery shopping for the same reason but do not do so communally (see Murphy 2001, 127–28). The good of community is achieved when agents have a shared goal and work together cooperatively for the sake of the community's good (what is sometimes called the common good; see Murphy 2001, 130–31).

Murphy argues that those working for the common good see it as intrinsically valuable and beneficial to the community as a whole (Murphy 2001, 128). Following Gregory Froelich, this sense of the common good has the rather obscure label of "good by way of causation" (Murphy 2001, 128). The root meaning of this phrase refers to something being a good both for a community (thus, a common good) as well as a good or an end for the individuals of that community (Froelich 1989, 48). Murphy's example of a good of this sort is a group of individuals who work to establish an art museum, both because of their own personal enjoyment of art, but also because they believe that an art museum is valuable for the entire community (Murphy 2001, 128). His example is tied to some of his basic human goods (e.g., aesthetic appreciation and excellence in agency) and might seem idiosyncratic insofar as its realization depends upon the interests of individual members of the community, but this need not be the case.[16]

Froelich argues that the natural world (which he calls the created world following Aquinas's theistic understanding of the world's origin) is a common good by way of causation (see Froelich 1989, 49–51). This is because of the value the natural world gains by sharing in the perfection of God who created it. Froelich's account implies that the value of the natural world rests on the truth of theism, which is in keeping with Murphy's conception of natural law.[17] If the natural world has intrinsic worth, the good of the natural world extends to and becomes an integral part of the good of the individual. Maintaining the good of the natural world becomes an end for individual human beings regardless of their individual desires and projects, simply because preserving such a good extends to and is integral to the entire community. This gives human beings a reason to care about the natural world. Insofar as human-induced climate change threatens the good of the natural world, a commitment to the common good provides natural law theorists such as Murphy with a reason to advocate for its mitigation.

Froelich discusses another way in which the common good could ground the requirement to alleviate climate change that does not depend upon the status of theism. He argues that for human beings, the common good is bound to the political community (by which he seems to mean organized social/civic communities). Froelich states:

> Since the political community is an ordered whole … its good consists in the preservation and tranquility of order. Such a good is not taken in opposition to the good of a single man [*sic*], for it is as a part of that order that man [*sic*] finds his highest natural perfection …. The political order, then, is a

> good for each citizen belonging to it. It is an intrinsic common good .... It is in fact man's [*sic*] highest natural end and for that reason takes precedence over his personal or private good.
>
> *Froelich 1989, 52–53*

Froelich is referring to the order of law, which he argues enables the attainment of virtue, which in turn perfects the individual members of the community and insures its stability (i.e., its order; see Froelich 1989, 53). But one could also argue that the political good is dependent upon the stability of the natural order as well, giving cities and countries reasons to pursue policies that promote sustainability and lessen human-induced climate change insofar as climate change threatens civic stability. On this understanding of the common good, the social order has as one of its fundamental ends the preservation and promotion of the good of the natural order.

Froelich's discussion of the role of virtue in the civic good raises questions about the place of virtue in the debate over what to do about climate change. His discussion also raises questions about the relationship between virtues and natural law theory (on this issue, see Murphy 2001, 212–19; and Finnis 2011, 90–91). My own view is that natural law theory describes a process of ethical judgment about how to pursue the ultimate end of human life. The virtues as dispositions toward actions that constitute or help to achieve the ultimate end work together with practical judgment in pursuing the ultimate end. Thus, virtues are an integral part of the good life, and practical reason and the virtues work together.

Recently, some climate scientists have argued that science alone is insufficient for addressing climate change; what is needed in addition to scientific research is the acquisition of virtues, such as love, humility, and wisdom (Hulme 2014, 300). Hulme argues that the exercise of virtue can benefit both the public debate over what to do about climate change and individuals in their own efforts to address climate change (Hulme 2014, 304). Thus, he ties the exercise of virtue to the common good and aligns both civic and individual aims in line with that good. Hulme argues that wisdom could enable us to evaluate competing claims and interests raised by climate research. Together with humility, wisdom could help us employ this research to address basic inequalities and vulnerabilities rather than relying on unproven technology to lessen the effects of climate change (Hulme 2014, 305). He cites climate scientist Tim Flannery's comments that without the virtue of love both for one another and for the Earth, equal to love of self, "no further progress is possible here on Earth" (Hulme 2014, 299). The development of such virtues could work to transform both individuals and their communities and enable the development of cooperative and effective policies designed to diminish the debilitating effects of climate change (Hulme 2014, 306). Hulme concludes:

> When we talk about climate change we should not start with the latest predictions from the climate models, nor whether we have passed some

> catastrophic tipping point; nor whether or not the Intergovernmental Panel on Climate Change should be trusted. We should start by thinking about what it means to be human. What is the good life and what therefore is an adequate response to climate change?
>
> *Hulme 2014, 308–309*

No doubt there are better and worse answers to these questions. For many in the United States, for example, the good life means a comfortable material existence. As I have argued in this chapter, questions about the good life and what it means to be human are central to natural law theory. Hulme acknowledges that such a conversation would be a paradigm shift from the usual way climate change is approached and would also be difficult and perhaps uncomfortable (see Hulme 2014, 309). But guided by the virtues, which together with the natural law enable human beings to work together for the common good, the hope would be that we could arrive at effective and sustainable actions to address climate change for the good of the community.

Conclusion

Natural law theory is an approach to ethics that involves identifying and protecting genuine goods that both enable human flourishing and are valuable in their own right. These goods provide human beings with reasons to pursue them or at the very least to safeguard them. Human-induced climate change constitutes a genuine threat both to the well-being of individuals and to larger entities such as the natural world and the common good that individuals have reason to care about for their own sakes. Thus, it follows on natural law theory that human beings ought to arrive at sustainable solutions to climate change.[18]

Notes

1 I do not pretend to have surveyed all natural law theories but have chosen to concentrate on what I take to be a representative sample. Furthermore, I do not defend or favor one account over the others; I focus instead on the resources that these theories bring to the discussion of climate change.

2 This is controversial, in part because of the is/ought controversy. For discussions of the role (or non-role) of human nature in natural law theory, see Crowe (2011, 299); George (1999, 83–87); and Donnally (2006, 22). Murphy argues that natural law theory requires a commitment to the claim that natural law is grounded in human nature: see Murphy (2001, 18, 137).

3 Murphy means by this the activities involved in choosing and acting well (see Murphy 2001, 114–18). It also involves what he calls "the state of having appropriate feelings and attitudes" (Murphy 2001, 114).

4 Two important caveats here. First, in late 2014 Chappell changed gender and name to Sophie Grace Chappell. Because the 1998 book I reference in this chapter was written by Timothy Chappell, her preference is that masculine pronouns are used while referring to the author of this text. Second, when the book I cite in this chapter was

published, Chappell was an adherent of a version of natural law theory, but she has since dropped that affiliation. Therefore, the reader should bear in mind that the label applies solely to the 1998 work I cite in this chapter.

5 Representative secondary literature includes Grisez et al. 1987; Lisska 1996; Murphy 2001; Porter, 2005; Rhonheimer 2000; Kluxen 1980; Hittinger 1987; Hall 1994; McInerny 1980; McInerny 1997, 40–62; Cromartie 1997; Goyette et al. 2004; Rziha 2009; and George 1999. Anthony Lisska has a critical survey of fairly recent literature on Aquinas's account of the natural law (see Lisska 2007).

6 Aquinas's account of wrongdoing is very complex. I provide an account of it in McCluskey (2017).

7 In his book on natural law, Murphy sets aside discussion of the role of religious authority in order to focus on the basic goods and his account of practical rationality (see Murphy 2001, 3–4). This raises the question of whether a religious commitment is in fact required for a natural law theory. Murphy affirms his view that religious authority is an important component of natural law theory but one that he was unable to develop satisfactorily in the 2001 project. Religion is also one of Murphy's basic goods, but as such, that gives people a reason to respect that good or at least not violate that good; it does not compel people to be religious.

8 At least as an intrinsically valuable good. Finnis discusses the common good but sees it as an instrumental good, which I discuss in note 15. I suspect that other natural law theorists would acknowledge the common good as a good but may not consider it to be a basic good. For example, Gómez-Lobo takes family and friendships as forms of community and discusses them as communal goods, but the basic good remains the family or the friendship (see Gómez-Lobo 2002, 14–16).

9 One rather extreme response to climate change, at least in my view, is the voluntary human extinction movement; for more on this idea, see Aufrecht 2017, 88–89, 100.

10 At this point, one might wonder about the Doctrine of Double Effect (DDE), which is accepted by some natural law theorists. DDE allows that one could perform an action that has among its foreseeable unavoidable effects an outcome that ordinarily would not be permissible, provided that certain restrictions are met (see Gómez-Lobo 2002, 79–81). Nevertheless, DDE does not sanction deliberate harm to a basic good even for the sake of a desirable end (on this issue, see Finnis 2011, 118–25). I thank Faith Glavey Pawl for raising this issue.

11 For a discussion of a sustainable agricultural program that includes the production of rice, see Shiva (1989).

12 Scott Davison argues that with only a few modifications, Murphy's account would support an environmental ethic. However, it is clear from Davison's correspondence with Murphy that Murphy would resist the major change that he would have to make, namely to admit that the good or flourishing of non-humans would constitute a reason for action (see Davison 2009, 5–7).

13 As Faith Glavey Pawl pointed out to me, the term "natural world" is ambiguous. It could refer to larger entities such as ecosystems or simply the set of all individual living organisms. On my view, these notions are interconnected insofar as individual organisms exist within and depend upon particular ecosystems. In what follows, I discuss the effects of climate change both on individual species and on ecosystems.

14 It also follows from these theories that human beings are profoundly dependent upon the welfare of the natural world for their own well-being.

15 Finnis also discusses the common good, but his understanding falls under what Murphy, following Froelich, would call the good of predication or the good of distribution (see

Murphy 2001, 127). Finnis defines the common good as a factor or set of factors that provide reasons for the participants to collaborate with one another (Finnis 2011, 154). This leaves open an instrumentalist account of common good. For this discussion, what is needed is a notion of common good where the cooperation in question is intrinsically valuable; thus, I will follow Murphy's and Froelich's accounts.

16 Aesthetic appreciation is a basic good for Murphy, but individuals are not obligated to pursue it. As long as they do not actively harm this basic good, they remain in good moral standing.

17 This would also give Murphy a reason to admit the natural world as a basic good, a move that he resists (see Davison 2009, 5–7).

18 I am grateful to Benjamin de Foy for his guidance and suggestions for this project and to Joe Salerno for his technical help. Ben Eggleston, Dale Miller, and Faith Glavey Pawl provided very helpful feedback for which I am also grateful.

Bibliography

Aufrecht, M. 2017. "Leave only Footprints? Reframing Climate Change, Environmental Stewardship, and Human Impact." *Ethics, Policy & Environment* 20 (1): 84–102. DOI: 10.1080/21550085.2017.1291823.

Aquinas, T. 1882-. *Summa theologiae*. In *Opera Omnia*, vol. 13–15. Rome: Commissio Leonina.

Berners-Lee, M., M. C. Hoolohan, H. Cammack, and C. N. Hewitt. 2012. "The Relative Greenhouse Gas Impacts of Realistic Dietary Choices." *Energy Policy* 43 (1): 184–90. DOI: 10.1016/j.enpol.2011.12.054.

Bryngelsson, D., Hedenus, D. J. A. Johansson, C. Azar, and S. Wirsenius. 2017. "How Do Dietary Choices Influence the Energy-System Cost of Stabilizing the Climate?" *Energies* 10 (2): article 182. DOI: 10.3390/en10020182.

Castañé, S. and A. Antón. 2017. "Assessment of the Nutritional Quality and Environmental Impact of Two Food Diets: A Mediterranean and a Vegan Diet." *Journal of Cleaner Production* 167: 929–37. DOI: 10.1016/j.jclepro.2017.04.121.

Chappell, T.D.J. (now Sophie Grace). 1998. *Understanding Human Goods: A Theory of Ethics*. Edinburgh: Edinburgh University Press.

Cromartie, M., ed. 1997. *A Preserving Grace: Protestants, Catholics, and Natural Law*. Grand Rapids, MI: William B. Eerdmans Publishing Company.

Crowe, J. 2011. "Natural Law Beyond Finnis." *Jurisprudence* 2 (2): 293–308.

Davison, S.A. 2009. "A Natural Law Based Environmental Ethic." *Ethics and the Environment* 14 (1): 1–13.

Donnally, B. 2006. "The Epistemic Connection between Nature and Value in New and Traditional Natural Law Theory." *Law and Philosophy* 25 (1): 1–29.

Emanuel, K. 2017. "Will Global Warming Make Hurricane Forecasting More Difficult?" *Bulletin of the American Meteorological Society* 98 (3): 495–501. DOI: 10.1175/BAMS-D-16–0134.1.

Finnis, J. 2011. *Natural Law and Natural Rights*. 2nd ed. Oxford: Oxford University Press.

Forseth, I., Jr, and A. F. Innis. 2004. "Kudzu (*Pueraria montana*): History, Physiology, and Ecology Combine to Make a Major Ecosystem Threat." *Critical Reviews in Plant Sciences* 23 (5): 401–13. DOI: 10.1080/07352680490505150.

Froelich, G. 1989. "The Equivocal Status of *Bonum Commune*." *New Scholasticism* 63 (1): 37–58.

Gardiner, S. M. 2010. "Ethics and Global Climate Change." In *Climate Ethics: Essential Readings* edited by S. M. Gardiner, S. Caney, D. Jamieson, and H. Shue, 3–35. Oxford: Oxford University Press.

George, R. P. 1999. *In Defense of Natural Law*. Oxford: Clarendon Press.

Gifford, R., and L.A. Comeau. 2011. "Message Framing Influences Perceived Climate Change Competence, Engagement, and Behavioral Intentions." *Global Environmental Change* 21 (4): 1301–7. DOI: 10.1016/j.gloenvcha.2011.06.004.

Gómez-Lobo, A. 2002. *Morality and the Human Goods*. Washington, D.C.: Georgetown University Press.

Goyette, J., M. S. Latkovic, and R. S. Meyers, eds. 2004. *St. Thomas Aquinas and the Natural Law Traditions: Contemporary Perspectives*. Washington, D.C.: Catholic University of America Press.

Grisez, G., J. Boyle, and J. Finnis, 1987. "Practical Principles, Moral Truth, and Ultimate Ends." *American Journal of Jurisprudence* 32 (1): 99–152.

Hall, P. 1994. *Narrative and the Natural Law: An Interpretation of Thomistic Ethics*. Notre Dame, IN: University of Notre Dame Press.

Henson, R. 2014. *The Thinking Person's Guide to Climate Change*. Boston, MA: American Meteorological Society.

Hittinger, R. 1987. *A Critique of the New Natural Law Theory*. Notre Dame, IN: University of Notre Dame Press.

Houghton, J. 2009. *Global Warming: the Complete Briefing*. 4th ed. Cambridge: Cambridge University Press.

Hulme, M. 2014. "Climate Change and Virtue: an Apologetic." *Humanities* 3 (3): 299–312. DOI: 10.3390/h3030299.

Karl, T. R., J. M. Melillo, and T. C. Peterson, eds. 2009. *Global Climate Change Impacts in the United States*. Cambridge: Cambridge University Press.

Kluxen, W. 1980. *Philosophische Ethik bei Thomas von Aquin*. Hamburg: Meiner.

Levine, L. 1993. "GAIA: Goddess and Idea." *BioSystems* 31 (1): 85–92. DOI: 10.1016/0303-2647(93)90035-b.

Lisska, A. J. 1996. *Aquinas's Theory of Natural Law: an Analytic Reconstruction*. Oxford: Clarendon Press.

_______. 2007. "On the Revival of Natural Law: Several Books from the Last Half-Decade." *American Catholic Philosophical Quarterly* 81 (4): 613–38.

McCluskey, C. 2017. *Thomas Aquinas on Moral Wrongdoing*. Cambridge: Cambridge University Press.

McInerny, R. 1997. *Ethica Thomistica: the Moral Philosophy of Thomas Aquinas*. Washington, D.C.: Catholic University of America Press.

_______. 1980. "The Principles of Natural Law." *American Journal of Jurisprudence* 25 (1): 1–15.

McMichael, A. J., R. E. Woodruff, and S. Hales. 2006. "Climate Change and Human Health: Present and Future Risks." *Lancet* 367: 859–69. DOI: 10.1016/S0140-6736(06)68079-3.

Murphy, M. 2001. *Natural Law and Practical Rationality*. Cambridge: Cambridge University Press.

_______. 2011. "The Natural Law Tradition in Ethics." *The Stanford Encyclopedia of Philosophy*, Winter edition. Edward N. Zalta (ed.). https://plato.stanford.edu/archives/win2011/entries/natural-law-ethics/.

Oo, A. Z., S. Sudo, K. Inubushi, M. Mano, A. Yamamoto, K. Ono, T. Osawa, S. Hayashida, P. K. Patra, Y. Terao, P. Elayakumar, K. Vanitha, C. Umamageswari, P. Jothimani, V. Ravi.

2018. "Methane and Nitrous Oxide Emissions from Conventional and Modified Rice Cultivation Systems in South India." *Agriculture, Ecosystems and Environment* 252 (1): 148–58. DOI: 10.1016/j.agee.2017.10.014.

Porter, J. 2005. *Nature as Reason: A Thomistic Theory of the Natural Law.* Grand Rapids, MI: William B. Eerdmans Publishing Company.

Rhonheimer, M. 2000. *Natural Law and Practical Reason: A Thomist View of Moral Autonomy.* Translated by G. Malsbary. New York: Fordham University Press.

Rottman, J., D. Kelemen, and L. Young. 2015. "Hindering Harm and Preserving Purity: How Can Moral Psychology Save the Planet?" *Philosophy Compass* 10 (2): 134–44. DOI: 10.1111/phc3.12195.

Rziha, J. 2009. *Perfecting Human Actions: St Thomas Aquinas on Human Participation in Eternal Law.* Washington, D.C.: Catholic University of America Press.

Shiva, V. 1989. *Staying Alive: Women, Ecology and Development.* London: Zed Books.

9

VIRTUE ETHICS AND CLIMATE CHANGE[1]

Sophie Grace Chappell

> Heart-breaking footage of a starving polar bear desperately searching for food on iceless land highlights the "real face of climate change," conservationists say. Photojournalist Paul Nicklen said his team were moved to tears as they captured the struggling animal's battle to stay alive on Canada's Baffin Island last summer. The "soul-crushing" scene, captured by marine conservationists SeaLegacy, shows the emaciated bear searching an abandoned Inuit camp for sustenance.
>
> "My entire SeaLegacy team was pushing through their tears and emotions while documenting this dying polar bear," explained Mr Nicklen.
>
> "It's a soul-crushing scene that still haunts me, but I know we need to share both the beautiful and the heart-breaking if we are going to break down the walls of apathy. This is what starvation looks like. The muscles atrophy. No energy. It's a slow, painful death. When scientists say polar bears will be extinct in the next 100 years, I think of the global population of 25,000 bears dying in this manner. There is no band-aid solution. There was no saving this individual bear. People think that we can put platforms in the ocean or we can feed the odd starving bear [but] the simple truth is this—if the Earth continues to warm, we will lose bears and entire polar ecosystems. This large male bear was not old, and he certainly died within hours or days of this moment. But there are solutions. We must reduce our carbon footprint, eat the right food, stop cutting down our forests, and begin putting the Earth—our home—first."
>
> *Molloy and Mills 2017*

Today the multiple and accelerating disaster of anthropogenic climate change confronts us with a swelling stream of stories as distressing as this one, or worse.

Climate change is a threat to charismatic megafauna like polar bears and whales and tigers; climate change is also a threat to smaller and less charismatic species that seem, nonetheless, to be crucial links in the overall flourishing of the global environment as we know it. For example, the Scottish midge (Latin name *Culicoides impunctatus*; Gaelic name *meanbh-chuileag*; Scots name *the wee basturts*), under threat from habitat loss as Scottish bogs dry out in a warming climate, is a key part of the food chain in the Highlands. It is preyed on by spiders and other predatory bugs, which, like midges themselves, are preyed on by small birds such as wheatears, skylarks, and meadow pipits, which are preyed on by buzzards and eagles and foxes. (The midge itself preys almost exclusively on climbers.) Climate change is a threat to ecosystems, such as the polar one. It is a threat also to the atmosphere, the envelope of breathable air no thicker proportionate to the planet than a layer of enamel is to a school globe, which occupies the first 6500 metres upwards from sea level anywhere on the Earth: increased heat melts Arctic permafrost, releasing more CO_2 into the air, which itself increases the Earth's heat. And climate change is a threat to humans: inter alia, it causes droughts that lead to desertification, rising seas that lead to floods, depopulation, famine, plague, political destabilisation, refugee crises, and war.

Like many other harrowing stories that I might have told, the story of the polar bear above produces responses in us that are very natural to talk about in the vocabulary of the virtues. This vocabulary is also, and not coincidentally, a vocabulary of the *emotions*, but I take the virtues to be, philosophically, the deeper phenomenon. Emotions are reactions, but virtues are—among other things—dispositions that underlie reactions, both in the sense that they produce them, and in the sense that they order, shape, and discipline them. So the virtues go deeper than the emotions, inasmuch as the virtues are part of what guarantees that there is such a thing as *right* emotional response. And that point is important when you consider how often environmental concern is dismissed as mere sentimentality. Against such dismissiveness, we may say that if there is such a thing as correct emotional response—as *virtue-guided* emotion—then there need be nothing wrong with founding at least some of our environmental thinking on such emotional response. As Nicklen says, "We need to share both the beautiful and the heart-breaking." Of course, a fully virtuous response will not *only* be a response to the tear-jerking. We might add to Nicklen's remarks that we also need to think about environmental damage that directly harms *nothing* beautiful, in *no* heart-breaking ways, and yet is environmental damage for all that; consider again the midge. But for the moment, I will stick with the dramatic cases.

Most obviously, the virtue-responses prompted by the story of the starving polar bear are pity and compassion: a sense that another creature is suffering, and that we, for all our differences from a polar bear, have some insight into what it would be like to suffer like that. Also that the creature is suffering gratuitously: the story may evoke something like righteous anger, a justice-based impulse towards rectification. If we put this impulse into words, we are likely to say that even if, as Paul Nicklen feared, nothing could be done for that individual polar bear, still something like justice needs to be done for all such creatures; and those who are responsible for the polar bears' plight—whoever they are, and if anyone is—need to be brought to justice, or at the very least made to change their ways.

Another part of our response to the story depends on our sense of reverence and awe for nature—a reverence that is very evident in so many of the nature programmes that we watch in the West, for instance Sir David Attenborough's *The Blue Planet*. This aspect of our response is not so much compassion for suffering, or anger at injustice, as outrage at defilement and spoliation. Despite much talk in academia about the disenchantment (*Entzauberung*) of nature in the modern era, it seems impossible to understand much of what the Green movement have had to say about pollution, environmental degradation, and species loss without taking literally what the Greens themselves often say about their campaigns and the outlook behind them. They say that they see the Earth as intrinsically precious, sacrosanct or maybe even sacred in and of itself (in the beautiful and pregnant words of a thoughtful observer from an earlier century, "I never saw an ugly thing in my life"[2]). And they say that this sense of the world precisely as *not* disenchanted, but as a rich and deep source of indefinitely many different kinds of enchantment, is the rationale for everything that they do.

So a broad spectrum of virtue-terms, and in particular compassion for suffering, a will to justice for all parts of the environment, and a sense of reverence for and protectiveness towards the enchanted in nature, will be useful to understanding philosophically how and why we respond as we do to cases like the starving polar bear. Alongside them, so will a broad spectrum of vice-terms. Indifference to suffering, lack of concern for justice, and irreverence towards nature are straightforwardly the opposites of the three dispositions of virtue that I have just identified, and it is pretty obvious why and how these count as dispositions of *vice*. But we can go further, by attempting to explain what drives these vicious dispositions; how they are rooted in or allied with other dispositions that also seem obviously vicious. Greed, selfishness, lack of imagination, and a brutish unresponsiveness to beauty are all dispositions we might mention here. So are apathy (which Paul Nicklen mentions), unwillingness to take responsibility and readiness to pass the buck to someone else, futility thinking, cowardice, conformism, and lack of moral vision, or of any sense of moral proportion. Again there is the vice of short-term thinking, which in the case of climate change, as indeed in some other cases, looks to be closely allied with the phenomenon of intergenerational theft. When you are in your seventies, like the current US president, why care what will happen in fifty years' time? Why not make a quick buck now before you die, and leave the youngsters to clear up?

Another vice that is slightly more subtle in form is what I shall call *false distancing*. False distancing is a vice that may well be in play when someone observes—truly—that the Earth has always been subject to fluctuations in temperature and in CO_2 and oxygen levels, and to new speciations and new extinctions. So, it is inferred, the changes that we are currently witnessing are novel only in that they are caused by one dominant species, ourselves. (And even that is not entirely novel, since most of our atmosphere's present oxygen content is there because of millions of years of plant life on Earth.) From all this it is supposed to follow that, for example, the extinction of the polar bear does not matter—not *really*, not *cosmically*, not *sub specie aeternitatis.* This poor argument seems best met by Bernard Williams' remark that "for most human purposes", eternity "is not a very good species to view [things] under" (Williams 1973: 118). A parallel argument concludes that death under torture does not matter; and if anything is a false conclusion, that is. No doubt there is some work to be done to find a philosophical and ethical balance between the urgent concerns of our own lives, and the universe's sublime detachment from all human projects and commitments. Still, the argument that climate change does not matter because it is just part of normal planetary processes is a non sequitur even if its premise is true. Further, the deployment of this sort of distancing move is, in practice, typically no more than an instance of bad faith and selective argument (which is why I call it *false* distancing, and a vice). People who say that "Climate change doesn't matter because cosmically *nothing* matters" are typically not philosophical-sceptical sages who live in barrels in the marketplace. Typically they say it in order to remove an obstacle that blocks them from pursuing something that actually they think matters very much indeed, such as making a quick buck.

This last vice of false distancing obviously has a cognitive or epistemic element to it. And alongside the moral vices just listed, we can also—these days I think we *must* also—mention some clearly epistemic vices that are likely to beset discussions of climate change and what to do about it. Living in denial is a vice; refusing to face up to the evidence, or to assess it objectively and fairly, is a vice; smearing whole intellectual enterprises—such as climate science, or indeed science itself—is a vice; the inverted epistemic snobbery that sneers that "the people of this country have had enough of experts"[3] is a vice (and, when the sneer comes from a clever man, is also an egregious case of *la trahison des clercs*). Today these epistemic vices are catching on. After all, some beliefs are more *convenient* than others. Obviously if there is no such thing as climate change then we don't have to do anything to fix what looks increasingly like a costly and difficult problem. Nowadays large swathes of US society in particular—these days, right up to the White House—have the air of a man who won't go to the doctor because he is half-subconsciously worried he might have cancer, and doesn't want to think about it, and inarticulately hopes that not thinking about the problem will make it go away. Alongside the virtues we have already mentioned, it takes temperance, courage, self-discipline, and humility—both in their moral and in their epistemic forms—to overcome this sort of self-serving ostrich mentality. (Or supposedly self-serving; in truth few things could be *worse* for our own interests, or at least for our common interests, than to ignore climate change.)

So far, then, virtue ethics looks like a moral theory that says the right things about the problem of climate change. For some virtue ethicists, all I need do to confirm this conclusion, and bring out virtue ethics as *the* moral theory that says the right things—that is, as the winner in the competition of the theories—is show that all the other moral theories do *not* say the right things about climate change. So, for instance, I might say that Hobbesian and Gauthierian contractarianism, and Rawlsian and Scanlonian contractualism, all do badly with environmental questions in general and climate change in particular, because these views by definition privilege moral agents, whom they see as contractors, over the world that, inter alia, they contract about. I might make a rather similar criticism of Kantianism, which divides everything there is into Ends In Themselves and Means To Those Ends, and seems to lack the resources to treat the environment or the climate as anything more than a Means. I might say that economic consequentialism struggles with environmental questions, because it is only interested in goods that can be priced so as to be entered into cost-benefit analyses; and environmental goods either have no definite price, which means cost-benefit analysis ignores them, or are of infinite value, which means cost-benefit analysis ignores everything else. Again, I might complain that welfarist consequentialism typically only considers the welfare *of sentient beings*, and so cannot consider the well-being of a non-sentient system such as the climate. (Sometimes welfarist consequentialists infer the non-considerability of non-sentients from the doctrine that all goodness is goodness *for* some sentient; there is no goodness that is not goodness for, no value without some sentient to experience it. This consequence then becomes either a *reductio* of the view, or a bullet for its proponents to bite.) We might, for example, if we are "deep Greens", call this view that non-sentients are morally inconsiderable an *intrinsic* fault of welfarist consequentialism, because there is such a thing as the climate's well-being, and welfarist consequentialism leaves it out. Or we might, for example, if we are "bright Greens", see it as an *instrumental* fault of welfarist consequentialism: whether or not there really is such a thing as the climate's well-being, welfarist consequentialism does less well in securing the good of sentient beings by reasoning as if there isn't. (A possible test case for these ideas: the New Zealand government's recent decision to grant "full rights of personhood" to the Whanganui River (*The Guardian* 2017). Is this treating the river *as having* those rights, or is it treating it *as if it had* them?) And so on.

I think there is something to every one of these criticisms that I have just listed. But I will not develop any of them here, for two reasons. First, because developing them would take me too far from my topic here, virtue ethics; it is for defenders of the moral theories criticised to respond to these criticisms. Second, because the question that we are now asking is "Which is the moral theory that says the right things about climate change?" I want to question this question.

Clearly it is a question with a certain chutzpah about it. It presupposes first that there is one, *exactly* one, moral theory that "says the right things about climate change"—where a moral theory is a systematised version of ethical thinking, usually alleged to possess a uniquely stringent rational and practical authority, that takes one and only one kind of thought or consideration to be crucial or basic or fundamental, ignores or side-lines or subordinates everything else to that, and builds or tries to build all our moral understanding, explanation, and justification around it and ultimately it alone. Second, it presupposes that we have some independent grasp on what those "right things" are. So at the very least the question is a nice illustration of moral theory's characteristically ambiguous position, halfway between two desiderata that seem to be in tension. We might call them the descriptive and the revisionary desiderata respectively. The one desideratum is that moral theory should exactly mirror, maybe even explain, what we already think are "the right things" to say about a given issue. The other is that moral theory should have the rational leverage to criticise, maybe even change, what we already think: should have the authority to reject some old and take up some new "right things" to say.

The danger posed by this descriptive/revisionary tension to the whole enterprise of moral theory is obvious. Many philosophers seem to find it comforting to respond to the descriptive/revisionary tension by appealing to the Rawlsian notion of reflective equilibrium. The trouble is that not just one but a multitude of different incompatible reflective equilibria seem equally possible, and equally defensible. Starting from any one theory that we may favour (call it T1), the danger is that T1 will mirror some of our extra-theoretical views but not others, yet claim credit for doing both. T1, we may be tempted to say, scores points by reflecting extra-theoretical views A and B. Yet T1 also scores points for *not* reflecting extra-theoretical views C and D; because T1 itself, since it generates the negations of C and D, shows that we shouldn't hold them.

If this is the game, then absolutely anyone can play it. Alongside T1 there can equally well be a T2, whose proponents have just the same right to say that T2's key merits are that it mirrors C and D, but discounts A and B, and explains why we should discount them. We seem to be heading towards a mere face-off between different sets of extra-theoretical views—A and B versus C and D; or at best, perhaps, towards a mere headcount of (or majority vote among) such views. If that's all we've got, then we ain't got much.

Myself, I suspect that larger-scale, messier, and more complicated versions of exactly this kind of impasse are only too prevalent in our practice of moral theory; that quite generally, a recurring question for the whole project of moral theory is "Why build on these extra-theoretical views and reject those, when you could equally well do the converse?" My own preferred strategy for getting round any such impasse is the same as it is with a literal mountaineering *mauvais pas*: retreat a little, descend a little, and have another look from a different angle. So forget about building a moral theory for a moment, and take another look at A, B, C, D, the extra-theoretical views that, in combination with our desire for a theory, got us into the impasse in the first place.

Notice that I have carefully called A, B, C, D "extra-theoretical views", and *not* called them "pre-theoretical views" or "intuitions". "Pre-theoretical" implies that these views are so to speak raw ingredients, waiting to be cooked up into something properly well-done and philosophical, namely a theory; which I think is the wrong way to think about such views. As for "intuitions", that word is even more freighted with implications I don't want, implications of unreflectiveness and primitive immediacy and knee-jerking gut-feeling-ness; implications too, at least to philosophers' ears, of a particular (and particularly suspect) quasi-perceptual moral epistemology. The words "pre-theoretical" and "intuition" are themselves part of what sets up the all-too-common antithesis—I say the *false* antithesis—between Naïve Pre-Philosophical Thinking and Sophisticated Philosophical Theory, and so helps lure us into the trap of thinking that those are our only alternatives, and that nothing but system-building moral-theoretical thought can be a properly reflective and philosophical response to the world's perplexities.

To the contrary, I suggest, much of the most interesting work that there is to be done in philosophical ethics is reflection on what we think "extra-theoretically"—that is, what we think when we actually let ourselves *think*, and are not busy trying to stuff what we think into the preordained and often Procrustean framework of some moral theory. When that happens, perhaps we will find all sorts of different considerations bearing on our moral thinking. Thoughts about principles, and consequences, and universalisability, and contracts, trusts and understandings, and virtues and emotions too—and other kinds of thoughts also: *all* of these will find a place in our moral understanding, though *none* of them is the only, or even the one uniquely-most-basic, moral currency in which we can deal. Maybe most of the really interesting, even most of the really sophisticated, stuff is of this extra-theoretical kind, and we should give up on the idea of systematic moral theory altogether. Maybe we should take the entire enterprise of philosophical ethics to be this sort of piecemeal and open-textured exploration of whatever we think most deeply true, and whatever we most care about.

As is probably clear already, this is my own picture of how philosophical ethics should be done. But doesn't this mean that I am not a virtue ethicist at all, but an anti-theorist? I have had positive things to say about virtue ethics as a promising approach to climate change. But I have just been advocating the rejection of the very idea of moral theory. It is hard to see how I can endorse the particular moral theory that is virtue ethics if I don't believe in *any* moral theory, in the sense that I have attacked—the sense of a systematised and monistic version of ethical thinking with a uniquely stringent rational and practical authority.

Now in one familiar usage of the phrase, "virtue ethics" is the name of a moral theory in precisely this sense. So for Rosalind Hursthouse, for example, virtue ethics is a third (genus of) moral theory alongside (the genera) deontology and consequentialism. We might borrow Hursthouse's puzzlingly urban-tribal metaphor and say that virtue ethics has gone through the stage of being "the new kid on the block, yet to establish its right to run with the big boys but not to be dismissed out of hand", and by now has "acquired full status" and is "recognised as a rival to deontological and utilitarian approaches, as interestingly and challengingly different from either as they are from each other" (Hursthouse 1999: 2).

"Full status" here pretty plainly means full status *as a moral theory*. Hursthouse does not much use the phrase "moral theory", but she does habitually write, as just above, about virtue ethics as another alternative exactly on a par with the alternatives consequentialism and deontology. So, presumably, virtue ethics for her is a moral theory in exactly the sense that they are both moral theories, with exactly the same standing as them to engage in what she seems, oddly, to envisage as gang warfare. (Or maybe not so oddly, in a sufficiently tough department.) This is plainest of all, perhaps, early on in her well-known article on abortion, where she sets up the three alternatives deontology, consequentialism, and "virtue theory" (as she there calls it) in strict parallel (Hursthouse 1991: 224–225):

> DP1.[4] An action is right iff it is in accordance with a moral rule or principle.
>
> CP1. An action is right iff it promotes the best consequences.
>
> VP1. An action is right iff it is what a virtuous agent would do in the circumstances.

DP1, CP1, and VP1 are the master-rules of deontology, consequentialism, and virtue ethics respectively. In my terms above, they specify the "kind of thought or consideration" that each of these three theories is going to take "to be crucial or basic or fundamental". Though Hursthouse does not actually say so, the three are clearly meant to be substantive and exclusive possibilities. She clearly intends a spelling-out each of these master-rules that avoids the apparently threatening consequences that each might turn out to be trivially true, and/or that someone who accepted one of them could painlessly accept one or both of the others also.

If all this can be secured—which can hardly be taken for granted, but I leave that aside here—then the next task is to specify a more precise content for these master-rules' definitions of right action. So Hursthouse gives us (1991: 224–225):

> DP2. A moral rule is one that … (i) is laid on us by God or (ii) is required by natural law [or] (iii) is laid on us by reason or (iv) is required by rationality or (v) would command universal rational acceptance or (vi) would be the object of choice of all rational beings [or (vii) … etc.].

> CP2. The best consequences are those in which happiness is maximised.
>
> VP2.[5] (a) A virtuous agent is one who acts virtuously, that is, one who has and exercises the virtues; (b) a virtue is a character trait a human being needs to flourish or live well.

No doubt further premises can be added to explain, in the case of deontology, which are the moral rules under whichever of specifications (i–vi) we go for; or in the case of consequentialism, which possible consequences are happiness-maximising; or in the case of virtue ethics, which the character-traits are that human beings need to flourish. But in that article it is not Hursthouse's main concern to add these further premises, even for virtue ethics. (She says more about this sort of question in her book.)

Virtue ethics in this sense, Hursthouse's sense, clearly is a moral theory in the sense in which, above, I suggested we should reject all moral theories. However, Hursthouse's is not the only possible sense for "virtue ethics". Given how long the idea of a virtue ethics had been around in modern academic philosophy before the 1990s (at least 50 years), and given that that idea has always been there in the western philosophical tradition, it is striking how long it took for Hursthouse's arrestingly neat and distinctive formulation of virtue ethics to appear. It is striking, too, how very unobvious it is that Hursthouse's predecessors in the virtue-ethics tradition, such as Plato or Aristotle or Aquinas in the ancient world, or Alasdair MacIntyre or Peter Geach or Elizabeth Anscombe or Bernard Williams or Charles Taylor or David Wiggins in the modern, would be happy to sign up for exactly her formulation. Perhaps more deeply and importantly, it is equally unobvious that they would be happy to agree with her evident belief that just her formulation, no more and no less, is what we need to situate virtue ethics as it should be situated, as a moral theory that competes for our assent with other moral theories like consequentialism and deontology.

If we share the doubts about the competition of the moral theories that I have described, then we could do something looser and less systematic than Hursthouse does with the idea of the virtues, while still remaining no less faithful than her to the tradition of the virtues as the writers just listed represent it. To put it roughly, we can move from a sufficiency claim to a necessity claim. Instead of saying, like Hursthouse and Philippa Foot, that a conception of the virtues is (at some level) *all we need* for the one theoretically adequate moral system, we may say, instead, that a conception of the virtues is *a necessary part* of any humanly adequate ethical outlook.

Though labels have no inherent importance, someone who takes this latter view might well, depending on the further detail of her views, still be worth calling a virtue ethicist. Even if she no longer avows a moral theory called virtue ethics, at any rate she has an ethics of the virtues. In fact I would argue, though for reasons of space I won't here, that it is only in this latter looser sense that Plato, Aristotle, Aquinas, MacIntyre, Geach, Anscombe, Williams, Taylor, and Wiggins are virtue ethicists. I would even argue, though again not here, that Aristotle, and Anscombe and Wiggins and Williams, too, are *both* virtue ethicists in this looser sense *and also* anti-theorists in the sense I defined above. The combination of anti-theory and an ethics of the virtues was possible for them, so no doubt it is possible for us too. (As indeed are other combinations: someone who in this sense has an ethics of the virtues, but is not a virtue ethicist in Hursthouse's exclusive sense, could also be a deontologist (like Onora O'Neill) or a utilitarian (like Roger Crisp), provided that too was not in an exclusive sense.)

This clarifies our understanding of what virtue ethics is. In fact, it tells us two different things that virtue ethics can be. One of them is a moral theory like Hursthouse's, which has at its heart a sufficiency claim: "Virtue is, in some fundamental sense, all you need to set up the true moral theory." The other is the less precise, and more capacious and flexible, thing that I am calling an ethics of the virtues or a virtue-ethical outlook, which centres on a necessity claim: "You cannot have an adequate ethical outlook without giving a central place to the virtues." I have argued that this is an important distinction. I have also argued that given the serious objections to the whole idea of a moral theory in the tight sense, it is philosophically preferable—and no less faithful to the tradition of the virtues—to adopt a virtue ethics of the looser ethical-outlook sort.

But either way—whether we consider virtue ethics as a full-blown moral theory, or more loosely as an ethical outlook—virtue ethics still faces two important objections which are especially clear when we think about environmental ethics, and in particular climate change. We can call these the timescale objection and the authority objection.

The authority objection is, roughly, the question "Says who?" That question of course arises for any and every ethical view whatever, and there might not seem to be much to be said about it beyond the usual platitudes (such as: "Well, *I* say so, with—I grant you—no special authority, but no special *lack* of authority either"). In fact I think there are interesting things to say about it that are fairly clearly proprietary to a virtue-ethical approach, to do with virtue ethics' rejection of the common modern idea that I have just alluded to, that ethics is essentially an area where no one has any special authority. To the contrary, virtue ethics insists that there are experts in ethics just as surely as in science—though perhaps rather more mysteriously. For the virtue ethicist, the views of the ethical expert, the *phronimos*, are pretty well definitive of ethical truth; if I cannot grasp her reasoning or agree with her verdict, that is not *her* problem, it is mine.

However, much of what I want to say about the distinctively virtue-ethical response to the authority objection, I have said or will say elsewhere (Chappell 2005; Chappell forthcoming). I therefore turn, for the remainder of this chapter, to the timescale objection.

The timescale objection says that virtue ethics works on the wrong timescale for dealing with urgent practical problems like climate change. Relative to climate change, what we need from philosophers is an urgent response articulating, detailing, and explaining philosophically what is "the right thing to do right now". We need philosophers to offer proposals about how to change our living, our thinking, our policies, and our framing of the problem—and we need them to be proposals that we can apply *right now* and that will make a difference *right now*.

Moreover, the state of the dialectic in philosophy today, as described above, makes it natural to suppose that what people are especially looking for from moral theories that discuss climate change (where I mean "moral theory" in the full-blown exclusive sense) is proposals *that are proprietary to their particular moral theories*. The thought is that, ideally (of course this ideal will mostly be only approached), proponents of moral theory T1 should be found offering a proposal P ("The right thing to do right now is …") where (a) P is immediately and obviously entailed by T1, (b) P strikes us as "the right thing to do" anyway (see above on how this might be problematic), and (c) P is *not* immediately or obviously entailed by any of T1's competitor-theories T2–TN. (If any or all of T2–TN actually entail the denial of P, then so much the better for the clarity of the dialectic.)

If this is the game—emergency ethics, as we might call it—then some moral theories seem better equipped to play it than others. Some theories, in fact, seem *designed* to play it. Most obviously, a certain form of consequentialism which is concerned "that things overall should go as well as possible" (Parfit 1987: 2) seems to be designed this way. This moral theory takes the notion of "things overall going as well as possible" to be determinate enough to yield extra-theoretically obvious results such as the badness of suffering, poverty, disease, and humans' and other species' extinction, and assumes therefore that we know enough to make recommendations about the right thing to do right now, which consequentialists can plausibly present as immediate consequences of their own moral theory, and perhaps not of anyone else's. It is no accident that the consequentialist Peter Singer is widely seen as the doyen of emergency ethics in this sense. Nor is it an accident that the "effective altruism" movement that has recently emerged in normative ethics is almost exclusively a consequentialist movement (a feature of it that effective altruists themselves, on their own principles, should presumably think regrettable).

By contrast, virtue ethics, on any proper understanding of it, seems very ill-equipped indeed to play the game of emergency ethics. Whether virtue ethics is a moral theory or the broader, less determinate thing that I call an ethical outlook, virtue ethics—it can plausibly be said—simply is not focused on urgent answers to urgent practical questions that take the form "The right thing to do right now is …".

Really, though? Why can't a virtue ethicist quite unproblematically come up with the kind of appeal to the vocabulary of the virtues that I explored at the start? Well, because the timescale objection suggests that actually such appeals are not unproblematic at all. The trouble with such appeals, according to the timescale objection, is that they misunderstand both what virtue ethics is all about, and also what it is to have a virtue.

For one thing, there are forms of virtue ethics, such as Michael Slote's in *From Morality To Virtue* (1995), that are founded on a contrast between the "aretaic" and the "deontic", the virtue-based and the obligation-based, and claim that virtue ethics is entirely concerned with the former and entirely uninterested in the latter. For such versions of virtue ethics, specifying "the right thing to do" is apparently *not even on the agenda.* And Slote's virtue ethics is not entirely exceptional in this. Even a far less anti-deontic philosopher like Hursthouse has her doubts about the notion of the right: "when a proponent of virtue ethics presents a criterion of morally right action, she does so 'under pressure' and 'only in order to maintain a fruitful dialogue with the overwhelming majority of modern moral philosophers'" (Hursthouse 1999: 69, as expounded by Hacker-Wright 2010: 223).

But even if virtue ethics does not go to Slote's extreme, it still does not look very focused on the agenda of emergency ethics. For whether "virtue ethics" means a moral theory or an ethical outlook, still virtue ethics is what the name implies: an ethics focused on the virtues. And that brings into sharp focus the question of what it is to have a virtue. Slote's exposition of the aretaic/deontic contrast may be overdone, but it does at least capture this important truth about virtue ethics: that virtue ethics is not primarily about what-should-we-do-right-now questions, it is primarily about *the dispositions that will make us into the kind of people who will give the right answers to* what-should-we-do-right-now questions.

The foundational texts of virtue ethics are, by general agreement, Plato's *Republic* and Aristotle's *Nicomachean Ethics*—and, I should want to add, Plato's *Protagoras.* All three of these works are not about emergency ethics; they are about moral education. It is not Aristotle's central aim in the *Nicomachean Ethics* (NE), nor Plato's in the *Republic* and *Protagoras*, to tell anyone how to act in a crisis. Plato's and Aristotle's target audience is an elite of present and/or future specialists and experts who (it is hoped or supposed) will or could be in a position to implement an educational policy in the *polis*; and their words are not mostly about crises or emergencies, but about habituation, inculcation, and training.

So in particular for Aristotle, the aim is not primarily or directly an ethics about "the right thing to do right now". Rather, his aim is to tell the upcoming generation of political leaders how to train, educate, and habituate the following generation of political leaders. That is why, if Aristotle has a single name for the subject matter of the *Ethics*, it more often seems to be *hê politikê tekhnê* or *epistêmê*, political craft or science, than *ta êthika* or *hê êthikê* (NE 1095a3, a16, 1099b30, 1102a8). It is why Aristotle takes *eudaimonia*, human flourishing, to be the objective not only of ethics but also of politics. It is why Aristotle is so familiarly pessimistic about the chances for character reform for anyone older than a certain age, and pessimistic too about the appropriateness, for anyone younger than that age, of directly ethical discussion of the kind found in the *Ethics*, as opposed to other less explicit forms of moral inculcation. ("It makes no small difference whether one has been habituated one way or the other from the very beginning of childhood; rather it makes all the difference, indeed every kind of difference there is," NE 1103b23; but "young men are not in their element listening to lectures on the political craft," NE 1095a2–3.) It is also for this reason that Aristotle is so often content to tell us no more about what particular actions are the right ones than *hōs dei*, "as it should be done" (see e.g. NE 1120a24–26); perhaps Aristotle might add that his concern as a scientist is with the universal, but every action is a particular. All of this body of reflection on what the virtues are and how to develop them is, as I say, presented by Aristotle as a prolegomenon not to private moral but to public political reflection: it is not for nothing that the last word of the *Nicomachean Ethics*, pointing us forward to the *Politics*, is *arxōmen*, "let us begin" (1181b24).

It follows from this conception of the ethical enterprise that for virtue ethics, in a sense, it is *always too late* for emergency ethics. To do right in the emergency confronting you, you need *already* to have developed the right dispositions—and most of that development needs to have happened years ago, in your immaturity. We are rather close here to Hegel's famous dictum about the Owl of Minerva. It is as if someone in a shipwreck had cried out "Bring me a plank right now!", and Aristotle had responded with a treatise on forestry.

What answer should this timescale objection get from defenders of virtue ethics (whether in the tight sense of a full-blown moral theory, or in some looser sense)? One preliminary contention: despite what Slote suggests, and at times even Hursthouse, they should not abandon the deontic. The rejection of the notion of the right by some modern virtue ethicists seems to me historically unjustifiable: virtue ethics today ought presumably to stand in some sort of lineal relation to the virtue tradition, but there is simply no basis at all for thinking that Platonist or Aristotelian or Thomist (or even Humean or Nietzschean) virtue ethics is interested only in the aretaic and not also in the deontic. Anyway, the move seems dialectically suicidal: it is not very hard to think up objections to a view of ethics that refuses to say anything about the right, talking only about (a variety of kinds of) the good.

The specific charge made by the timescale objection is that virtue ethics is not well-placed to deal with questions of emergency ethics like climate change, because (as I put it above) virtue ethics is not about what-should-we-do-right-now questions, it is about the dispositions that will make us into the kind of people who will give the right answers to what-should-we-do-right-now questions.

This charge has some bite against both virtue ethics *taken as a moral theory*, and virtue ethics *taken as an ethical outlook*; but the arguments turn out differently in the two cases. The charge means that virtue ethics can have nothing distinctive and proprietary to say about our responses to emergency-ethical issues like climate change: it will make some recommendations in the vocabulary of the virtues, and others in other vocabularies. That is a big problem for virtue ethics as a moral theory, because that version of virtue ethics typically insists that, at least at some level,[6] it can provide a distinctive and proprietary version of moral deliberation. (Recall its sufficiency claim, as I called it above.) It is less of a problem for virtue ethics as an ethical outlook, since there is nothing in this looser form of virtue ethics to disallow the possibility of using, as of course we do use, a whole variety of considerations in deliberation. (Recall the necessity claim.) The vocabulary of the virtues is a key part of this variety, but not the only part.

Still, for both forms of virtue ethics, the key worry raised by the timescale objection remains. The timescale objection prompts the worry that exhortations to virtue will always come too late. For the virtues come *slowly*: they appear in people's characters only after decades of habituation. So it can look like exhorting people to do the virtuous thing here and now is always going to be advice that is either unneeded, because those who *have* the virtue will act on it anyway; or impossible to follow, because those who *do not have* the virtue cannot act on it. Since acting virtuously involves not just doing the right thing, but doing the right thing for the right reasons, they need to have grown the virtue first. And that takes years; years that, for them, are no longer accessible because they are already in the past.

No doubt the dilemma is suspiciously neat, like Plato's dilemmas about knowledge and ignorance in the *Meno* and the *Theaetetus.* But as with those dilemmas, there are interesting things to be learned by thinking how to solve it. The natural suggestion is that the resolution depends on the intermediate states, and on the dynamics that get us from one to the other. In this case that means the states between lacking and having virtue (or vice), and the process of learning virtue (/vice). In real life, no one is either completely virtuous or completely lacking in virtue: in that sense, both sides of the supposed dilemma are unrealistic. Hence adjuring people to do the virtuous thing is not like addressing them in a language that they don't speak at all; it is more like improving their grasp of a language that they already speak a bit, by practising it with them and getting them to repeat and imitate the sounds. So exhortations to others that are expressed in the vocabulary of the virtues need not involve us in saying, in effect, "Go back in time and spend the last ten years becoming *this* sort of person—and then you'll be able to respond properly to the present crisis." It is more like an invitation to see things a certain way, and to respond in a certain way, that is open and possible to them right now: a way that may not be *full* virtue, but is at least an approach to it, and an improvement on anything else that they would be at all likely to do instead. Ethical learning goes on throughout our lives, and it is unfortunate that Aristotle says so much to suggest that there is a hard and fast contrast between learning the virtues and possessing them in their full forms; even if, with a laudable kind of inconsistency, he also has plenty to say about practice, repetition, and imitation as means of moral learning.

Might we still be worried, despite these thoughts, about the difficulty of someone "acting as the *phronimos* acts" when it takes years to become a *phronimos*, and when, for whatever reason, that someone has not put in the years? If we are, two further thoughts may also be helpful. The first is that someone who is not really a *phronimos* might nonetheless succeed, more or less, in imitating the *phronimos* in some particular performance. For sure, her action will be a mere imitation of the *phronimos*'s performance: she will not—and without the history cannot—act *exactly* as the *phronimos* does. Still, she does manage an *imitation* of what the *phronimos* does. That is something. And something is better than nothing.

This brings us to my second "further thought", which takes up a well-known idea of John McDowell's (1995: 74). Through the sort of experience that McDowell tantalisingly calls "something like conversion" somebody, he tells us, "might suddenly or gradually become *as if* he had been properly brought up". Such an experience might be a stand-in for part or whole of a "proper" moral education, rather in the way that Donald Davidson's Swampman has an internal set-up that enables him to copy exactly the referential practices of those who have the right causal background for achieving genuine reference—even though, magically emerging, as he did, fully-formed from his swamp, he does not share that background.

For McDowell this is, apparently, a borderline possibility, a conceivable-in-principle way in which external reasons to act well might be generated in someone who lacks the right background to get them the usual way—by habituation. If like me you take it to be a central part of our ethical life that values can and do confront us pretty well directly (in some good sense of that difficult word), in the pivotal ethical experiences that I call epiphanies, then you may want to treat this possibility of "conversion" as much less borderline and merely-possible, and much more central and really-actual.

In another ethical vocabulary, Aquinas's, one might also talk at this point of the distinction between habituated and infused virtues (*Summa Theologiae* 1a2ae.63.3). The infused virtues provide another way round the timescale objection, because they involve the very same dispositions and actions as the habituated virtues—and yet they arise in us immediately and without habituation, by the direct influence upon us of God's grace.

Notes

1 The author is grateful for the help and comments from audiences in St Andrews and the Open University, Andy Holland, and the volume's editors.
2 John Constable, quoted in Thomas (1983: 70).
3 Michael Gove, in an interview with Faisal Islam of Sky News on 3 June 2016: https://www.youtube.com/watch?v=GGgiGtJk7MA
4 Here I have tidied up Hursthouse's notation a little. She puts "P.i.", with a Roman "i", for all three of these first premises, then switches to Arabic "2" for the three second premises. I have added a letter before the P to indicate which theory each premise is a premise of, and regularised the numeration to Arabic.
5 More tidying up: what Hursthouse labels "P. i a." and "P.2", I label "VP2(a)" and "VP2(b)" respectively.

6 "At least at some level": it is possible to cast virtue ethics as an indirect moral theory, i.e. as one in which agents' conscious-level, real-time deliberations need not take the theory's own distinctive and proprietary form—though their not having that form must have some justification from deeper down in the theory. For virtue ethics as an indirect moral theory, deliberative thoughts that *exemplify* virtue X need not be deliberative thoughts that *mention* virtue X. This is Hursthouse's well-known thesis about "v-thoughts" and "v-rules" (Hursthouse 1999: 30), and it gives a way for the moral theory of virtue ethics to accommodate the objection that, as a matter of familiar experience, deliberation does not in fact proceed exclusively through the vocabulary of the virtues. It also opens up this version of virtue ethics to all the usual objections to any indirect moral theory, in particular the alienation and self-effacement objections.

Bibliography

Chappell, Sophie Grace. forthcoming. *Epiphanies: An Ethics of Experience.*

Chappell, Timothy. 2005. The good man is the measure of all things': objectivity without world-centredness in Aristotle's moral epistemology. In: Gill, Christopher ed. *Virtue, Norms, and Objectivity: Issues in Ancient and Modern Ethics.* Oxford, UK: Oxford University Press, pp. 233–255.

Hacker-Wright, John. 2010. "Virtue ethics without right action". *The Journal of Value Inquiry* 44, 209–224.

Hursthouse, Rosalind. 1991. "Virtue theory and abortion", *Philosophy and Public Affairs* 20.4, 223–246.

Hursthouse, Rosalind. 1999. *On Virtue Ethics.* Oxford: OUP.

McDowell, John. 1995. "Might there be external reasons?" In J. E. J. Altham and Ross Harrison eds. *World, Mind and Ethics: Essays on the Ethical Philosophy of Bernard Williams.* Cambridge: CUP, pp. 68–85.

Molloy, Mark and Emma Mills. 2017. "Dying polar bear roams iceless land for food in 'soul-crushing' video." *The Telegraph* 11 December 2017.

Parfit, Derek. 1987. *Reasons and Persons.* Oxford: OUP.

Roy, Eleanor A. 2017. "New Zealand river granted same legal rights as human being." *The Guardian* 16 March 2017. https://www.theguardian.com/world/2017/mar/16/new-zealand-river-granted-same-legal-rights-as-human-being.

Slote, Michael. 1995. From Morality to Virtue. Oxford: Oxford University Press.

Thomas, Keith. 1983. *Man and the Natural World.* London: Allen Lane.

Williams, Bernard. 1973. "A critique of utilitarianism". In Bernard Williams and J.J.C. Smart eds. *Utilitarianism: For and Against.* Cambridge: CUP.

10

FROM CARING TO COUNTER-CONSUMPTION

Feminist Moral Perspectives on Consumerism and Climate Change

Regina Cochrane

Introduction: Climate Change, Global Justice, and Consumerism

Alluding to the dire warnings issued in the October 8, 2018 Special Report released by the United Nations' Intergovernmental Panel on Climate Change (IPCC), the UK's *The Guardian* announced: "We have only 12 years to limit climate change catastrophe." The IPCC insists that urgent changes are needed to keep global warming to a maximum of 1.5°C and thus to prevent hundreds of millions of the Earth's inhabitants from being subjected to worsening risks of extreme heat, floods, drought, and poverty. Without such changes as reforestation, a shift to electric transportation systems, and increasing use of carbon capture technology, a rise of 2°C could, among other extremely serious issues, double the number of people subjected to water stress and halve the number of habitats supporting crop-pollinating insects (Watts 2018). However, while the IPCC is generally acknowledged as the "premier authority" on climate change—due largely to the extensive body of scientific expertise that it has mobilized (Harris 2016, 18–19, 27)—and therefore officially warranted to certify its consensus on global warming as "unequivocal" (Garvey 2008, 16–17, 91), "we need more than that if we want to act on the basis of those [scientific] facts. The something more which is needed involves [moral] values" (1–2).

What we decide to do about climate change depends ultimately upon the kind of moral values we hold; these values lead, in turn, to the type of reasoned ethical principles to which we commit ourselves and the right conduct which they prescribe (Garvey 2008, 2, 33–34). Given that such ethical principles are necessarily based on logical consistency and, thus, fairness (45), they stipulate that benefits and burdens be distributed equally among people, provided no morally relevant reasons for unequal distribution exist (42, 67–69). Moreover, because climate change impacts certain regions of the world and specific groups of people disproportionately, it must be approached as a matter of distributive justice (76). As Paul G. Harris notes:

> Climate change is a profound matter of injustice imposed on people who are already poorly off and who usually have no say in the matter. The adverse effects of climate change most harm the weakest and poorest countries and people of the world, imposing burdens on those states and those people least responsible for causing it, most exposed to it and most vulnerable to its ravages (for example, sea-level rise and storm surges, drought and floods, disease and heat stress) and least able to pay for mitigation and adaptation. These effects are mostly a result of what other, mostly affluent, people have done in places far from where the pain is felt. Consequently, because of its causes and its consequences, climate change is an issue that cries out for justice.
>
> *Harris 2016, 37–38*

The distributive aspect entailed in climate justice is, however, considerably more complicated than this. While most commentators assign the blame for climate change to the developed nations of the global North (Garvey 2008, 73–76), Harris also targets the growing number of "new consumers" from the developing countries of the global South. Adopting high-consumption lifestyles similar—or even superior—to those in industrialized countries (Garvey 2008, 6), he considers such newly affluent populations, with their "unjust *luxury* emissions" which "free ride" on their states' limited obligations (153), to be the "new drivers of climate change" (129). Moreover, those negatively impacted by climate change also include certain people from the global North. The most seriously affected victims of Hurricane Katrina also tended to be among the most systematically marginalized populations in New Orleans: the residents of poor neighborhoods, members of racialized communities (especially African-Americans), and the elderly. In fact, many of these people ended up becoming trapped—and some even died—in the inundated city (Cannavò 2008, 179–180). Women burdened with traditional caregiving responsibilities, subjected to heightened risks of sexual assault during emergency situations, and/or afflicted by poverty, can also be rendered particularly vulnerable by climate disasters, together with their children (Gaard 2015, 23). As well, certain LGBTQ persons, in both North and South, can be similarly put at risk by forces such as the religious right, which reportedly attributed Hurricane Katrina to an act of divine retribution targeting sexual minorities (24).

Returning to the affluent of the global North and South, such populations are widely considered to have causal responsibility for climate change because their rampant consumption is using up vastly more than their fair share—not only of the world's resources but also and more significantly—of the world's finite carbon sinks (Garvey 2008, 72–73, 76). While all humans need to support their basic requirements for food, shelter, heat, and other necessities (69) by releasing subsistence emissions into the atmosphere, the global middle-class and elites are also spewing out enormous amounts of luxury emissions (81) that cannot be absorbed and thus end up heating the planet (69). Hence, due to their "massive, foreseeable, and avoidable wave of voluntary" consumerism (Harris 2016, 140), the world's privileged have "frivolous[ly]" (17) "polluted the atmosphere, often as a result of conspicuous consumption and other activities that are not essential to life or happiness" (3). Indeed, one mind-boggling example of a major contributor to carbon gas emissions is the rapidly rising consumption of meat which now releases more greenhouse gases than the entire transportation industry (20). Moreover, much of this increase is supported by the "new consumers," approximately half of whom live in the global South (114), who are emitting greenhouse gases "at a pace and scale never experienced in human history" (124). Harris, in fact, bluntly attributes such quintessentially consumerist behavior to the strong encouragement provided "by global capitalists seeking to benefit from increasing buying power in the developing world" (132) via capitalizing on "the limitless emotional desires of the world's affluent" classes (181). Therefore, even for individualist adherents of liberal democratic capitalism, like Harris, ultimately "there is a grudging acceptance of capitalism's destructive action," albeit made by "pinning full responsibility on the agency of definite individual consumers" (Luke 2008, 148).

When examining which groups are directly impacted by the contemporary climate crisis and which ones are directly complicit with it, the poor of both South and North—and, especially, the female, racialized/sexual minority, and elderly poor—are likely acknowledged as belonging to the former group and privileged, middle-to-upper-class men—of both North and South—as part of the latter. However, to what extent middle-class women are complicit with climate change is presently subject to greater debate. Indeed, feminists themselves hold conflicting positions on the involvement of relatively privileged women due to their complicity with consumerism-fueled global warming. Moreover, what type of position they take tends to depend upon the type of feminist theory to which they subscribe and, more specifically, the form of moral theory which it advocates. With respect to the particular issue of consumer-driven climate change, the most common feminist moral responses involve various forms of care ethics. Hence, in addressing the climate crisis, cultural feminists take inspiration from a "maternal feminist" care ethics while socialist feminists—who see privileged women as complicit in consumerism—tend to adopt a post-consumerist alternative rooted in feminist and socialist moral values which may also be conjoined with a more collectivist version of care ethics.

To make the case that debates about feminist care ethics and moral values can contribute to possible solutions to climate change, the chapter is organized as follows. First, I will examine arguments for and critiques of maternal feminist ethics, and how women's vulnerability to climate change and purported virtue in resisting consumerism tend to play out in climate emergencies. Next, I will review the dramatically different arguments made by socialist-feminist philosopher Kate Soper for a positive and non-moralistic "alternative hedonism," which she ultimately sees as compatible with a form of care ethics, together with her more recent extension of this work to climate change. Finally, I will conclude with some suggestions, by feminists and others, about the need for an ecological citizenship that could help address limitations in Soper's "alternative hedonism."

Consumption-Fueled Warming, Maternal Care Ethics, and Women's "Virtue"

Although it has been traced back to the "matriarchal vision" of certain first-wave feminists (Donovan 2012, 31–62), contemporary cultural feminism is generally viewed as originating from a split within the 1970s American radical feminist movement (Echols 1987), which the former has largely but not entirely replaced. While cultural feminists share radical feminists' rooting of women's oppression in patriarchy—a system oriented around male dominance and power (Tong and Fernandes Botts 2017, 2) and men's control of women's sexuality, reproduction, and self-esteem (40)—they have discarded radical feminism's androgynous ideal for a woman-centered perspective (44) celebrating motherhood (3). Hence, despite women's ethnic, class, and personal differences, cultural feminists insist that their "traditional culture," values, domestic practice, and caring labor are "nearly universa[l]" and thus have led "to the formation of a particularly female epistemology and ethic" (Donovan 2012, 167–168). This is a "gynocentric" ethic that is life-affirming and non-imperialistic and evinces a fundamental respect for the concrete details of everyday life and the natural environment (170). As such, it rejects patriarchal society's downgrading of caring labor in favor of reason and objectivity (Ruddick 1999, 407) and thus "resist[s] the lure of abstraction" for "the particular knowledge that women acquire from their suffering of and resistance to oppression" (408). In fact, "strong" cultural feminists tend to insist that "women's characteristics and values are for the good, indeed are superior and ethically prior to men's, and should be upheld" (Evans 1995, 76; Tong and Fernandes Botts 2017, 44; Ruddick 1999, 405). The main proviso here is, however, that "mothering" is not confined to biological parenting but extended to all types of caretaking, maternal roles (Donovan 2012, 171) and thus that it is possible for men to learn from such maternal values too (61).

Although often taken as synonymous with feminist ethics *per se* (Jaggar 1991, 83), feminist care ethics constitutes—together with an alternative, status-focused feminist approach—a gender-centered rethinking of traditional ethical theory's neglect of women's reality or its failure to adequately address it (Tong and Williams, 2019). At a more theoretical level, the knowledge underlying women's "maternal thinking" is grounded in a feminist standpoint epistemology—an "engaged vision of the world opposed and superior to dominant ways of thinking … [which] is produced by the experience and … distinctive [caring] work of women" (Ruddick 1999, 406). Moreover, based on such experience, moral psychologist

Carol Gilligan claimed, in her influential text *In a Different Voice* (1982), that males and females in patriarchal societies develop gender-linked values and virtues. Critiquing the "universal," six-stage process of moral development outlined by educational psychologist Lawrence Kohlberg, Gilligan found that women typically remained at stage three (Tong and Fernandes Botts 2017, 183–184). This is the level at which, for Gilligan, the individual balances caring for others with maintaining care for the self (Puka 1993, 216). While men usually moved on to stages four or five (Tong and Wilson 2019), which encompass a form of moral reasoning relying on abstract principles of fairness, rights, and justice, women's affinity for relationships and connection consequently fosters what is often misunderstood as a more limited form of moral thinking centering on particular people's concrete needs and responsibilities (Tong and Fernandes Botts 2017, 183–184). Although Gilligan's original thesis concluded that men's style of moral reasoning was simply different—and not superior—to women's, in her subsequent work she "hinted that the ideal moral thinker might after all be more inclined to an ethics of care than an ethics of justice" (185).

Closely following up on Gilligan's approach, philosopher of education Nel Noddings declared that care is most likely foundational to ethics and justice is merely its corrective (Noddings 1984, 42–46, 94–98; Tong and Fernandes Botts 2017, 186–187). Therefore, care is not simply different from but considerably better than the ethics of justice (Noddings 1984, 83). "[T]raditional curriculum [i]s a masculine project, designed to detach the child from a world of relation and to project him, as object, into a thoroughly objectified world," asserted Noddings. This is in marked contradiction to "an alternative feminine approach [which] allows the child to remain in relation and also to grow intellectually" (Noddings 1984, 192). Such a change involves an "increase in the spirit of caring—that spirit that many refer to as 'the maternal attitude'" (197). Hence, the paradigmatic relationship for maternal care ethics becomes, with its ultimately problematic political subtext, the mother-child dyad.

In direct contrast to a Kantian-type set of rationally chosen moral principles that are impartial, universalizable, and that can be used to determine the correctness of an action, maternal care ethics can thus be classified as a form of contextual moral theory (Tronto 1993, 248–249):

> In any contextual moral theory, morality must be situated concretely, that is, for particular actors in a particular society. It cannot be understood by the recitation of principles. By this account, morality is embedded in the norms of a given society. Furthermore, contextual moral theory directs attention away from the morality of single acts to the broader moral capacity of actors. To be moral is to possess a moral character, or, … virtue …. Thus, morality cannot be determined by posing hypothetical moral dilemmas or by asserting moral principles. Rather, one's moral imagination, character, and actions must respond to the complexity of a given situation.
>
> *Tronto 1993, 248*

Consequently, there is a similarity between maternal care ethics and other, more mainstream examples of contextual ethics like Aristotelian—and, in its more Catholic form, Thomistic—virtue ethics and the Scottish Enlightenment's notion of "moral sentiments" (248). In keeping with this are the common complaints, by certain feminist critics, that cultural feminism has something of a religious flavor. Moreover, it is consistent as well with cultural feminist Josephine Donovan's linking of maternal ethics to Adam Smith's liberal classic, *The Theory of Moral Sentiments* (Donovan 2007, 181, 186)—which is to be strongly contrasted with neoliberalism (Brown 2015, 92–94, 102–103).

Surveying the existing literature on gender and climate change, Seema Arora-Jonsson finds two recurring themes—women's vulnerability and women's virtuousness vis-à-vis the environment. More specifically, "women in the South are [seen as] extremely vulnerable to climate change while women in the North are much more conscientious when it comes to dealing with climate change, possessing virtues of environmentalism" (Arora-Jonsson 2011, 744). Although Arora-Jonsson makes no mention of feminist moral approaches to climate change in her subsequent analysis of these dual themes, it is clear that she considers maternal care ethics to be the predominant orientation in feminist climate change literature. This is likely due to the strong association between environmental feminism/ecofeminism and cultural feminism. Moreover, notes Arora-Jonsson, it is "in large part driven by the desire to put women and unequal gender relations on the map in relation to discussions on climate change" (747). Hence, in the language of maternalist cultural feminism *á la* Nel Noddings, the vulnerable woman of the South becomes the "cared-for" and the virtuous woman of the North the "one-caring" (Noddings 1984, 4).

The themes of vulnerability, virtue, and maternal care ethics are, to cite just one particularly illustrative example, front and center in Meike Spitzner's chapter, "How Global Warming is Gendered: A View from the EU."

> [G]lobal warming is not only predominantly brought about by the global North, it is overwhelmingly an effect of decisions and actions, made by men occupying positions in institutions set up and staffed by men.

Elaborating on what she sees as the gender differences in perception of social and environmental issues, she contends:

> Now surveys show that women and men perceive social and environmental risks differently. Women's higher sensitivity to risk, to social and environmental qualities, and their competencies in caring labor, lead them to view market-based international climate policies less positively than men.

Spitzner observes of the "virtuous" women of the global North:

> [R]esearch shows that women in general are more able than most men to recognize the urgency of global warming and to adjust their personal energy consumption, for instance by changing shopping habits.

Indeed, she notes, making a veiled reference to women of the global South:

> [T]he gender ascribed caring economy would demonstrate its lower climate change impact. ... [For] [w]omen worldwide … [a]s climate change sets in train its harsh conditions and new scarcities threaten vulnerable groups, women's unfairly large (unpaid economic) workload will increase. As things stand, 70 per cent of the world's poor are women, mostly unsupported mothers and elderly, lacking any social and economic capacity to adapt to unanticipated environmental impacts.
>
> *Spitzner 2009*, 218–219, 222–223

Arora-Jonsson is highly critical of this focus on women's vulnerability and virtue vis-à-vis climate change for its three basic assumptions which she rejects as simplistic. First, the notion that women of the global South are vulnerable because they are the poorest of the poor she finds to be homogenizing. In actual fact, the often-repeated "assertion that women make up 70% of the poor"—which Spitzner cited directly above—is, she argues, "anecdotal" rather than based on statistical or empirical evidence. Not all women of the global South are poor but the rising incomes of middle-to-upper class women do not necessarily protect them from violence caused by gender discrimination. Second, the claim that women from the South experience an increased mortality rate during natural disasters depends upon such disasters making existing forms of discrimination against women worse, often as a result of multiple, intersecting systems of oppression such as class, ethnicity, and caste (Arora-Jonsson 2011, 746). Third, the assertion that women from the global North are more environmentally aware—and thus virtuous relative to their male counterparts—seems to be due to a tendency to preconceive such women as pro-environmentalist (747). Such vulnerability and virtue thus end up representing women's roles and the differences between men and women as static (748).

Retracing the arguments presented by Arora-Jonsson—albeit at a more in-depth, theoretical level —Bernadette Resurrección adds a different focus and some key terminology to critiquing notions of women's vulnerability and virtue in the face of climate change. Evoking overt references to cultural feminism, she labels such views as "essentialis[t] in … assigning gender differences as innate and transcultural properties" (Resurrección 2013, 34). Moreover, taking issue with "essentialist linkages" between women and nature, she finds assumptions about women's "fixed caretaker roles" to be "disassociated from wider relationships and webs of power" (35). Consequently, the "claim to an essential feminine subject tied to nature homogenizes other women subjects, blurring the possibility of more context-specific subjectivities rooted in class, ethnicity, age, eco-zones, and so on and denying … where positive opportunities may also inadvertently lie" (41).

Indeed, based on their essentialist arguments about Southern women's vulnerability and Northern women's virtue, cultural feminists blatantly end up assigning the greater part of the responsibility for global warming to over-consuming men of the North. Given the homogenizing gender stereotypes and anecdotal evidence marshalled to support this conclusion, however, it is worth examining briefly what has given rise to this consumerism and then how it is gendered. In the transition from pre-modernity to modernity, Aristotelian ethical notions of human nature rooted in finite desires gave way to an understanding of the human as a social animal with infinite desires (Sassatelli 2007, 36). Moreover, these infinite desires and the new needs to which they gave rise (20) emanated from the creative fantasies of the mind rather than the sensorial practices of the body, like eating and drinking (17). Central to this process was the emergence of the new mode of capitalist production which was induced by an ascetic, Protestant ethic favoring hard work and capital accumulation.

However, equally fundamental to this gendering of consumerism, contends Roberta Sassatelli, was a new mode of consumption (10) that created the market needed to supply large amounts of goods at accessible prices (16). This form of consumption responded to the status aspirations of the newly emergent bourgeoisie (15) by "rationalizing" them via the dynamics of fashion, thus maintaining the wheels of production in constant motion (23). The lifestyle promoted by modern capitalist society—initially for the upper-middle classes/upper bourgeoisie—was therefore a hybrid of hedonism and asceticism, "a high culture of waste and refinement and a bourgeois one of thrift and prudence" (22). Yet with the further development of capitalism, which has culminated in the contemporary neoliberal era, this lifestyle has trickled down to the middle- and lower-middle classes, leaving behind its tempering association with asceticism and leading instead to a fully hedonistic ethic of consumerism (40, 50).

Extending and gendering elements of Sassatelli's analysis as developed above, contemporary neoliberal consumerism can be understood as comprising separate "male" and "female" moments. Given men's more extensive involvement in the higher echelons of the global capitalist system, as critiqued by Spitzner, they do have significantly higher economic resources, on the *average*, than do women. However, the men who form part of this higher echelon are also involved in "rationally" investing much of this capital in businesses that encourage the hedonistic consumerism of the middle- and lower-middle classes, thus reducing their own consumerism accordingly. Hence, essentialist notions of consumer-fueled climate change, applied to men of the global North, are probably overestimated. Moreover, similarly essentialist ideas of the "virtuous" women of the global North likely lead to underestimating their consumerism and, in turn, contribution to global warming. In fact, there are specifically gendered styles of consumption to which women are attracted and in which they become complicit, as Sassatelli explains, drawing on the ideas first outlined by sociologist Georg Simmel:

> [F]ashion is often associated with women and femininity. This association has clear historical and cultural roots: the historical weakness of their social

> position oriented women towards "comparatively greater uniformity" and fashion was instrumental in enabling them to express themselves through a shared language, becoming "the valve through which a woman's craving for some measure of conspicuousness and individual prominence finds vent, when its satisfaction is denied her in other fields[.]" … *[G]ender* is a social construction which shapes practices of consumption at the same time as it itself is constructed by such practices.
>
> *Sassatelli 2007, 28*

Although it is important to understand women's complicity in such consumerism as a reaction to their lack of power, it is also vital, from a feminist perspective, to "more capaciously recognize the conservative nature of some gendered consumption in the attempt to push beyond it" (Littler 2009, 183). This does not imply that women's carbon emissions are exactly equal to men's—in either the North or among the affluent new consumers of the South—but that the differences between "gender-marked" emissions *in* the North are much less significant than those *between* total emissions from the North compared with those from the South.

While the essentialism underlying cultural feminism and its maternal care ethics clearly limits its potential to adequately respond to consumer-caused climate change, there are a host of other related issues here. As Judith Evans notes:

> This approach is problematic even if we accept its emphasis on culture. It ignores differences among groups of women, or is unable to address them adequately. It faces the danger that female virtue can be used against women, and has been. It may merely echo women as described by patriarchy, may simply invert the dualisms we know. Further, feminists like these are often regarded as elitists who "know" what women are, should be, can be, should say: who know what "woman" can be. And there remains the issue of who decides what bona fide female experience is.
>
> *Evans 1995, 78*

Evans' critique resonates with complaints made by Alison Jaggar that "feminine ideals are often subversive of feminist aspirations" (Jaggar 1991, 92). Sarah Hoagland objects, among other things, to the "unidirectional description of caring" promoted by maternal ethics (Hoagland 1991, 249), in which "the one cared-for cannot understand what the one-caring needs or wants" (251) and the "trust the one cared-for is supposed to place in the authority" of the one-caring (252). Finally, as Marilyn Friedman warns, given that "care and justice overlap" necessarily, "morally adequate care involves considerations of justice" (Friedman 1993, 259). Consequently, when "untempered by justice …, care degenerates precipitously" (267).

There is one final critique of maternal care ethics that follows directly from Friedman's argument about the need to supplement care with justice. This is something that comes into play with the more political and collectivist versions of care ethics, espoused by socialist and even some liberal feminists—the issue of institutionalization:

> In an account of care's progressive struggle with sexism, [Gilligan's] level 3 care might be faulted for its lack of political sense or institutional focus out of which a sense of solidarity with other women and a need for cooperative social action might derive. Care's almost total lack of social-institutional focus at level 3 certainly raises questions about its general moral adequacy. The attempt to balance serving others with self-care at level 3 does not solve the problem of slavishness. It merely tempers and accommodates to it in a morally questionable way. This accommodation is then intellectualized, especially in Gilligan's description of level 3, by portraying it as a necessary complement to "male-oriented" justice.... By contrast, a truly liberated ethic for women (and other oppressed groups) might speak in a truly new voice, expressing themes of unfolding, liberated experience.
>
> *Puka 1993, 220*

Feminism, Alternative Hedonism, and the Pleasures of Sustainable Consumption

Socialist feminism is a synthesis of an androgyny-affirming radical feminism (Tong and Fernandes Botts 2017, 40) with a more New-Left-inflected version of Marxism. As such, it promotes a "both/and" alternative to, rather than a cultural feminist gynocentric inversion of, hierarchical, male-centered dualisms like reason/emotion and culture/nature. Moreover, it roots women's oppression in a systemic "capitalist patriarchy" (Jaggar 1988, 124), that incorporates within it an interconnected set of intersectional dominations focusing on racism, imperialism, heterosexism, disability, and other related forms of subjugation (Gordon Jaggar 1988). Therefore, male domination is not only grounded in the patriarchal system of domestic relations but also in the capitalist economic foundation of society (Jaggar 1988, 159). Its abolition will therefore require a transformation of the entire political economy, including the political economy of gender and sexuality (147). Such a historical transformation (128) will, for socialist feminists, entail "the full development of human potentialities for free sexual expression, for freely bearing children and for freely rearing them" (131). Moreover, to realize this goal, it will demand the elimination of the division of labor by sex and thus the end of socially constituted categories of male and female roles (132). Consequently, the whole social order of capitalist patriarchy (329), with its ensuing enmity between male and female, is not inherently permanent (330) but can be historically transcended.

Given that Soper's perspective is unorthodox and the "alternate hedonism" she proposes, as a response to climate change, is the outcome of decades of her own work, it is appropriate to begin by situating her proposal in her earlier writing of the 1990s and her own reading of socialist feminism. Taking issue, in the introductory pages of *Troubled Pleasures: Writings on Politics, Gender and Hedonism*, with both "'maternal' and [postmodern] 'difference' feminisms," she aligns herself with the "phenomenolog[ical] and existentialist … wing of feminist argument." Moreover, although convinced that we must break with capitalism and institute a more egalitarian distribution of wealth and resources, she casts doubt on Marxist projections like the future promise of "indefinite abundance" and the working class as the "ultimate" agent of liberation (Soper 1990, 9). Indeed, framing her overall project as "a 'politics of needs' [interrogating] … what is needed as a condition of human flourishing and happiness" (46), she takes strong issue with "communist-capitalist complicity … around needs" (47) and, especially, with Marxists who were attempting to smash capitalism via "the more peaceful 'consumer Communist' route to revolution" (49). Such consumerism is being increasingly challenged by a "growing ecological awareness of the contradiction between the pursuit of prosperity as currently conceived and the survival of the planet" (47), she insists.

> It is for this reason that we may now need as socialists to entertain the idea of a more trans-class, consensual contestation of a free-market system which in coming up against the absolute limits of ecological resources is also beginning to deny everyone's needs.
>
> *Soper 1990, 65*

Moving back from the ecological to the social justice implications of need, Soper calls as well for a fundamental questioning of affluence that would foster a "significant reduction of First World demands upon the impoverished sectors of the globe, and hence a realistic chance of inaugurating less exploitative North-South relations" (66).

Both poverty in the global South, which is the outcome of "of a century and a half (at least) of aggressively imperialist consumption" (Soper 1990, 33), and global environmental issues are, for Soper, integrally related to capitalist consumerism. "We live in a society that offers so little diversification in the way of work and creative activity that consumption comes to figure as the main vehicle of self-expression" (81). It is her hope that, by "dwel[ling] less on the expansion of consumer wealth and more on the pleasures of curbing that expansion" (68), we can inspire people through a focus on the hedonistic "pleasures an alternative society might afford in compensation for a more restrained material consumption" (69). She elaborates, in more depth, on this notion as follows:

> The aim is not deprivation but a different balance of gratification: to become abstemious in the consumption of material commodities, but more profligate to compensate in the so-called "goods of the spirit." Community, friendship, sexual love, conviviality, wild space, music, theatre, reading and conversation, fresh air and uncontaminated land and water: it is with these attractions that

> we should be cultivating desire and pandering to the senses, rather than with images of improving shares, the flight to the Bahamas and the second car. We should aspire to a new eroticism of consumption, an altered aesthetic of needs: one which makes the senses recoil from commodities which waste the land, throw up ugly environments, pollute the atmosphere, absorb large quantities of energy and leave a debris of junk in their wake.
>
> *Soper 1990, 33–34*

The compensations for curbing consumerism would be significant, argues Soper, including "[l]ess boring work, less time spent on it, less pollution, more space" (Soper 1990, 81–82). Indeed, it could also entail:

> people becom[ing] more exorbitant in their demand for such goods as to walk where they want to, when they want to; to loiter talking on street corners; to travel slowly; to have solitude, space to play and time to be idle.
>
> *Soper 1995, 269*

As a result, "many ... would if given the chance opt for these conditions of self-realization at the cost of an increase in more tangible gratifications" (Soper 1990, 82). In fact, she insists that such compensations be an integral part of any realistic socialist vision:

> [T]he [socialist] appeal to altruism has to be complemented by an appeal to self-interest, where what is stressed it not simply the risk and misery to be alleviated, but the pleasures to be realized by breaking with the current market-defined and capitalist promoted conceptions of the good life.... [This] will require the socialist argument to be backed ... by an alternative hedonist vision: by very different conceptions of consumption and human welfare from those promoted under capitalism, pursued under "actual existing socialism" or hitherto associated with orthodox socialist theory.
>
> *Soper 1995, 271*

The kind of socialism that Soper supports, therefore, requires that socialists "become advocates of an alternative utopia of wants" (Soper 1995, 272) and thus be not only humanists and feminists but also hedonists (Soper 1990, 69).

What are the implications of Soper's "alternative hedonism" for feminism? In opposition to cultural feminists, Soper's targets consumerism—rather than men, their over-consumption, and "patriarchal" economic/political institutions—as the main cause of climate change. Moreover, rather than seeing women's "virtues" and maternal ethic as fostering a solution, she notes how "movements for sexual emancipation have been co-opted by the market." Indeed, casting aside women's continuing history of counter-consumerist activism, the contemporary "feminization of consumption ... celebrate[s] the license given to self-making, gender performance and the re-construction of identity by consumer culture." Taking direct issue with such current feminist trends, Soper presents her "alternative hedonism" as "the impulse behind a new 'political imaginary' that could help to … fulfil some of the more radical aspirations that have been associated with the movements for gender and sexual emancipation" (Soper 2009A, 93). Indeed, moving beyond a "work ethic and culture rooted in the conventional gender division of labor" (96) would be conducive to feminist aspirations for much needed changes in the domestic realm:

> [T]he provision of a decent but not extravagant standard of living, based on part-time employment for everyone … would be consistent … with some quite drastic and arguably very beneficial changes in the conduct of domestic life, the upbringing of children and human relationships. Domestic and nurturing tasks could be shared between both sexes, allowing both a richer experience and more mutual understanding. Domestic work itself could become less constraining and tedious because there would be time and energy to provide ourselves with things and services we currently have no choice to buy or hire. There would be time to spend on each other, to think and talk, teach and learn, make and create, and simply to be idle.
>
> *Soper 1990,* 35–36

Moreover, the slower pace of life and reduction in stress would also contribute to feminist goals such as improved health (Soper 2009A, 95), "new modes of thinking about human fulfilment and the life-work balance" (96), and "a general rethinking of pleasure and the good life" (98) that encompasses the need for "both sensual pleasure and more spiritual forms of wellbeing" (95).

Returning to the theme of "alternate hedonism" during the opening decades of the new millennium, Soper reframes her environmental focus to zero in on climate change, adds some reflections on the effects of neoliberalism, and delves, in more depth, into economic changes required to introduce a post-consumerism future. "Consumerism is the major cause of global warming and wrecking the planet for future generations," she declares, launching her 2017 New System Project paper, "A New Hedonism: A Post-Consumerism Vision" (Soper 2017, 5). Rejecting mainstream approaches to climate change, based on technological fixes, she advocates the "alternative hedonist" view "that even if consumerism were indefinitely sustainable it would not enhance human well-being" (20–21). Given "the oppressive effects of the climatic impact of First World affluence" on the rest of the globe, she asks, why should such wasteful consumption practices still "remain exempt from the kinds of criticism that we now expect to be brought against racist, sexist or blatantly undemocratic attitudes and modes of behavior" (46)? Soper connects the rise of neoliberalism to the loss of more progressive thinking about work, consumption, and use of time and to fears about precarious employment and the growing "gig economy" (35). As regards economic measures that could facilitate a sustainable "alternative hedonism," she seems particularly interested in the implementation of a UBI (universal basic income), which seems to be "gaining traction around the world," including "projects [that] are planned for Ontario, Canada" (38). Also mentioned with favor are projects such as LETS (Local Exchange and Trading Scheme), carpooling and recycling networks, anti-sweatshops campaigns, and activism around animal testing and factory farming (38–41). Such a "commitment to an alternative politics of prosperity based on a sustainable economic order" she views "as comparable in the forms of social transformation and personal epiphany it will demand to those brought about through the feminist, anti-racist and anti-colonialist movements of recent history" (46).

The most significant change introduced by Soper in her more recent work, in fact, involves a deepening of her treatment of the ethical underpinnings of her hedonistic approach. This involves reiterating the basic claim she made earlier; that is, that over-consumption, by the global North, "is unlikely to be checked in the absence of a seductive alternative conception of what it is to flourish." The citizen-subject she evokes here, she emphasizes, is more complex than the appetite-driven individual of neo-classical liberalism. Thus, the "attractants for well-to-do consumers themselves" will necessarily involve motivating factors encompassing new forms of desire and "new modes of thinking about human pleasure and self-realization" (Soper 2009B, 3–4), such as the novel proposal of "time affluence" (6). However, added to this is the notion that "the more selfishly motivated preference" for the hedonistic aspects of consuming differently are seen as "inseparable from a more collectively oriented concern to avoid contributing to ... danger, pollution, and congestion" and thus ultimately climate change (5).

Hence, Soper now argues that there is "a considerable overlap between alternative hedonist types of motivation and the altruism of the green or ethical consumer." Both "engage in a distinctively moral form of self-pleasuring and in a self-interested form of altruism which takes pleasure in committing to a more socially accountable mode of consuming" (Soper 2009B, 5). In other words, Soper's hedonistic ethic of counter-consumption is now conjoined with a collective form of care ethics which derives pleasure from moral behavior. She describes this ethic as encompassing "a cosmopolitan[1] care for the well-being of the more deprived people of the world and a concern about the quality of life of future generations" (Soper 2009a, 99; Soper 2017, 46) whose lives will be marked—and very possibly severely limited—by climate change. Soper recognizes that such an approach will denote a break from orthodox left analyses (Soper 2009B, 6), given that it challenges the Marxist priority of producing over consuming. Thus, it questions not only who owns the means of production—and thus the primacy of the working class's revolutionary agency—but also the key environmental issue of what is being produced and which methods are used to produce it (9).

However, while Soper assumes that implementing a UBI would encourage people to adopt an "alternate hedonism" leading to a reduction in consumerism and carbon emissions, critics tell a different story. John Clarke, of the Ontario

Coalition Against Poverty, considers the pilot project run in Ontario—which Soper mentioned in her 2017 paper—to be "fundamentally a neoliberal trap." Rather than reversing the trend toward precarious, low-waged work, it would, in fact, expand it. As its original right-wing proponents—including Milton Friedman—effectively insisted, UBI "must not augment but, rather, replace existing systems of social provision, so that people become shoppers in the neoliberal marketplace, picking their way through the privatized rubble of the social infrastructure." In fact, the signs were especially ominous during the Ontario pilot, given that clients, previously on social assistance, were due to lose counselling and referral services and benefits for supports like special dietary needs, medical transportation, glasses, hearing aids, mobility devices, and even service dogs (Clarke 2018).

> It is necessary to say very bluntly that the notion that the state is going to provide a level of secure and unconditional income that removes economic coercion from the dealings between employers and workers is just not rooted in reality.
>
> *Clarke 2018*

Supporting Clarke's conclusion, Daniel Zamora adds: "UBI isn't an alternative to neoliberalism, but an ideological capitulation to it … [which] would universalize precarious labor." Certainly, "[n]o existing economy can pay for a generous basic income without defunding everything else." Rather than a UBI, "a universal job guarantee and a reduction in work hours still represent the most important objectives" (Zamora 2017). In fact, this latter demand is much more consistent with Soper's "alternative hedonism."

Moreover, Soper's assumption that, stimulated by a "changed perception of the 'good life' (an 'alternative hedonism')," a "broa[d] trans-class body of concerned producers and consumers" could instigate the "pressure … to secure the necessary policy changes" for a "democratically achieved process of change in the West today" (Soper 2008, 199) seems rather naïve. Indeed, this idea directly contravenes Wendy Brown's assertion that neoliberalism is resulting in a drastic "hollowing out of contemporary liberal democracy" (Brown 2015, 18). In other words, "commitments to equality, liberty, inclusion, and constitutionalism" are being subordinated, under neoliberalism, to economic growth (26), "evacuating democratic principles, eroding democratic institutions and eviscerating the democratic imaginary" (28) and transmuting democratic self-rule[2] into management (20). Whereas Soper's "alternative hedonism" calls for a rethinking of the "good life," neoliberal rationality eradicates any such notions (43) by transforming human beings into forms of "human capital" (65). Leaving behind "all concern with justice" (128)—and echoing Soper's comments on mainstream approaches to climate change—"the aim is to produce and implement practical solutions for technically defined problems" (129) and to stifle the capacity to critique and contest (208). As such, "citizen virtue is reworked as responsibilized entrepreneurialism and self-investment" (210). The nation state thus ends up being reconceived along "the model of Wal-Mart, where managers are 'team leaders,' workers are 'junior associates,' and consumers are 'guests'—each integrated into the smooth functioning of the whole and bound to its ends" (211).

Conclusion: Toward a Feminist Ecological Citizenship and a Sustainable Consumption

Summing up her critique of the virtue-and-vulnerability stance espoused by cultural feminist advocates of maternal care ethics, Arora-Jonsson reiterates her position on the two-fold role of gender in the climate crisis. "Unequal gender relations do not cause or aggravate climate change. But gender relations do determine how the environment is managed" (Arora-Jonsson 2011, 750). While she rejects the essentialist argument that the domination of women causes or aggravates global warming, Arora-Jonsson does believe that feminism has a vital role to play in addressing the power relations producing women's vulnerability during situations of climate emergency. Socialist-feminist Kate Soper would definitely concur, insisting that patriarchy is not implicated in causing or exacerbating climate problems but that women can be rendered particularly vulnerable during disasters linked to rising global temperatures. She would be especially adamant, as well, about affluent women's "virtue," given their complicity in the consumerism-fueling carbon emissions. Moving beyond Arora-Jonsson's critique, Soper looks instead to the social proposals espoused by the women's movement and how they could be fostered by embracing the seductive pleasures of an ecological hedonism and reinforced by a cosmopolitan care ethic concerned about the poor and future generations.

Cultural feminism's maternal care ethics has also been challenged for its lack of political and institutional focus; in contrast, Soper does suggest that proponents of her alternative hedonism "pressure the state" to implement the democratic changes needed to reduce consumerism and thus global warming. Yet, Soper's treatment of how exactly this trans-class coalition of "producers and consumers" might collectively act to "pressure the state" and halt climate change is extremely undertheorized. Moreover, even if the details of this political engagement were to be discussed in greater depth, this raises a more complex question. Will politics still function as in liberal times, given that what we have now is now largely a *neoliberal state* rather than a *liberal democratic one*? Would the state still respond to "pressure" exerted by citizens, given that democratic principles and justice are now being subordinated to economic growth? Can an "alternate hedonism," oriented around moral values, ethical principles, and right behavior and inspired by notions of the "good life," be meaningfully re-cast as "a practical solution for a technical problem"? How will people view caring, justice, and ecological virtues when citizenship is redefined as "responsible entrepreneurship and self-investment"? How will we work toward social and ecological change in the precarious, non-inclusive, and geoengineered Republic of Walmart, where political critique will be merely dismissed as inefficient and disruptive?

Given that the climate crisis is happening concurrently with—and being significantly accelerated by—the political crisis of liberal democracy, fighting climate change also will necessarily require challenging what Brown critiques as neoliberalism's "substantive disembowelment" of democracy (Brown 2015, 9). Therefore, for feminists responding to climate change in neoliberal times, this would involve the political quest for an appropriate form of feminist ecological citizenship. Some efforts have already been made to develop such an orientation. Chris Cuomo embarks on this path by declaring that:

> individual and household reductions in greenhouse-gas pollution will be effective only if they are deep and widespread, and only if they are accompanied by meta-level efforts, but meta-level policies and corporate practices seem unlikely to emerge without significant support from "below."
>
> *Cuomo 2011, 708*

Though she likely espouses a liberal feminist perspective, Cuomo's approach would be compatible with Soper's trans-class appeal and, furthermore, Cuomo's anti-essentialist, collectivist stance on care, which appeals to women to "effectively act on their caring while also making it contagious through the creation of a more effective political will," resonates with aspects of Soper's "alternate hedonism" and her cosmopolitan care ethic (Cuomo 2011, 708). However, Cuomo's "political will," like Soper's, is still oblivious to the neoliberal erosion of the democratic state. In contrast, rejecting both the "fixed and private feminized identities that are themselves depoliticized (for example, maternalism)" as well as "the depoliticizing forces of neoliberalism … and environmental managerialism" (MacGregor 2014, 630), Sherilyn MacGregor seems, at first glance, to move significantly beyond both Soper and Cuomo. Yet, in the final analysis, it seems that her main problem with neoliberalism is its reliance on "incontestable scientific truths," "inevitabilities," and "consensus" (629)—that is, with its discourses of Truth—which contain dissent and "eras[e] social difference." In the name of affirming such difference,[3] MacGregor thus takes a stance in direct opposition to Soper's "left-green prescriptions for sustainable development," (630) which, she implies, seek to "turn back time" by returning society to a "steady state." Instead, she argues for a "principled stance for climate justice and a radically *different* future" (626) (emphasis added).

In an era marked by the double crises of climate and democracy, the project of a feminist ecological citizenship oriented around an "alternative hedonism" clearly merits further work and consideration. Added to this task, as well, is the urgent need to counter right-populist movements, which are arising around the world in conjunction with the neoliberal challenge to democracy. Moreover, this is a development that is also co-opting feminist environmentalism itself (Cochrane 2007), as it is a social movement theory which, in an age of neoliberalism, displaces—rather than supplements—issues of economic inequality with notions of cultural diversity (Cochrane 2014, 593–594). We live in complex times and the obstacles blocking our movement toward a feminist ecological citizenship are many. Hence, it may be worth considering quickly some "modest" advice offered, as a "navigational compass" that can "serve to guide society's general direction" (Geus 2003, 89–91), by green political theorist and advocate of a "sustainable hedonism" very much in keeping with Soper's: Marius de Geus.

Taking as his point of departure the dominant moralistic and hedonistic conceptions of "the good life"—that is, living in an ethical, just, and responsible way, as advocated by Plato and Aristotle, versus pursuing need satisfaction, happiness, and pleasure, as promoted by Hobbes and Locke (Geus 2009, 113–115)—Geus charts a third path between these two extremes, which takes the form of a "sustainable hedonis[t]" alternative (119). Opting for a pragmatic resolution that moves beyond these conflicting versions of "the good life," he defends a liberal acceptance of lifestyle diversity and freedom, as allowing for ecologically accountable lifestyles to take various forms. This plurality of life choices he then conjoins with a moralistic insistence on responsibility, as consistent with protecting the vulnerable and future generations and decreasing material consumption (120). A multiplicity of ways of life should be encouraged, insists Geus, as long as appropriate ecological constraints are maintained—that is, provided that individuals restrict their "ecological footprints"[4] to sustainable limits (124).

Attaining such a "sustainable hedonism" will also depend, Geus adds, upon cultivating the "greened" moral disposition that arises from practicing the "ecological virtues." Among such green virtues, he includes "prudence, respect, care, moderation, self-control, tolerance, balance of activities and awareness of limits to growth and material welfare" (Geus 2009, 125–126). Moreover, practicing such virtues will, he emphasizes, necessarily involve "a permanent learning process that seeks to integrate individual freedom and diversity of lifestyles with social-ecological and moral responsibilities towards our fellow humans and descendants" (128) and towards non-human nature (Geus 2003, 47, 75). In keeping with the ideas of Soper, Geus's virtuous "sustainable hedonism" would be based on a life of mindful consumption, "restrained luxury" (179), "pleasant frugality" (198), and "comfortable austerity" (88), where life would be "outwardly simple and inwardly rich" (Dominguez and Robin as cited in Geus 2003, 195) and nature would be unharmed (Geus 2003, 182). As such, it would qualify as another form of "alternative hedonism which is both enjoyable and rewarding without exceeding our fair share of the earth's resources" (Geus 2009, 128).

Kate Soper's "alternate hedonism" has much to contribute to a *seductive*, feminist, counter-consumerist response to climate change. Adding to this, Geus stresses the need for "more profound political steps and social measures" (Geus 2003, 131) that could encourage citizens to critically reflect and intensively deliberate on issues requiring "deep social, political, economic, and cultural changes" (132). However, Geus's proposed "pragmatic" restructuring of citizenship ultimately boils down to a more decentralized and less bureaucratic approach (70) to the liberal nation state (75) and to its conjoined free market economy, albeit one that is stringently ecologically limited (70). Hence, like Soper's alternative, Geus's "sustainable hedonism" fails to take into account neoliberalism's fundamental undermining of democracy. In order for this issue to be effectively addressed, the critical reflection and deliberation Geus calls for must go well beyond issues of over-consumption and climate change to a reconsideration of liberal democracy itself. Therefore, Soper's feminist response to climate change—like Geus's allied position—needs to move beyond a *seductive* counter-consumerist response to climate change to a *seductive* re-imagining of democracy and of a feminist ecological citizenship.[5]

Notes

1 There are some interesting overlaps between Soper's approach and that of Harris. Not only do both embrace forms of cosmopolitanism but Harris makes several statements that resonate with Soper's "alternative hedonism." See, for example, his discussion on p. 178 of his book (Harris 2016).

2 In keeping with Brown, Soper argues that "none of [the classical liberals] presented self-interest as the sole motive of human action" (Soper 2008, 192). Hence, she shares much of Brown's defense of liberalism vis-à-vis neoliberalism. See Brown (2015, Chapter 2).

3 For a critique of postmodern arguments, like MacGregor's, defending "difference" as the alternative to universalism per se, see my paper "Campaigning Against 'Fundamentalist' Globalization" (Cochrane 2011).

4 Drawing on Mathis Wackernagel and William Rees, Geus defines individuals' "ecological footprint" as "a measure of the land area needed to sustain the levels of resources consumption and waste discharge" they produce. This "fair Earthshare" of ecologically productive land amounts to approximately 1.8 hectares per person (Geus 2009, 124–125).

5 Soper does provide a hint about how her approach can be further developed. "Consumerism's general tendency, in fact, is to offer material acquisition as a means of satisfying desires requiring altogether more complex intellectual and emotional gratification" (Soper 2017, 17), she notes. What she alludes to here is the deeper problem motivating consumerism—the problem of alienation. (For a discussion of alienation, see: Jaggar 1988, 307–317.) Hence, Soper's "alternative hedonism" needs to be extended beyond our alienation from work, society, sexuality, and nature to our alienation from politics. From a socialist feminist perspective, this could also include questioning the hierarchical politics underlying not only Stalinism but also Leninism, which has influenced even more moderate forms of socialism, like social democracy. For an anti-Leninist perspective on socialism, which also moves considerably beyond the limits of Brown's conception of liberal democracy, see Miguel Abensour (2011).

Bibliography

Abensour, Miguel. 2011. *Democracy against the State: Marx and the Machiavellian Moment.* Translated by Max Blechman and Martin Breaugh. Cambridge, UK and Malden, MA: Polity Press.

Arora-Jonsson, Seema. 2011. "Virtue and Vulnerability: Discourses on Women, Gender and Climate Change." *Global Environmental Change* 21, no. 2 (May): 744–751. https://doi.org/10.1016/j.gloenvcha.2011.01.005.

Brown, Wendy. 2015. *Undoing the Demos: Neoliberalism's Stealth Revolution*. Brooklyn, NY: Zone Books.

Cannavò, Peter F. 2008. "In the Wake of Katrina: Climate Change and the Coming Crisis of Displacement." In *Political Theory and Global Climate Change*, edited by Steven Vanderheiden, 177–200. Cambridge, MA and London: The MIT Press.

Clarke, John. 2018. "Basic Income: Progressive Cloak and Neoliberal Dagger." *The Bullet,* April 4. https://socialistproject.ca/2018/04/basic-income-progressive-cloak-and-neoliberal-dagger.

Cochrane, Regina. 2011. "Campaigning Against 'Fundamentalist' Globalization: Toward a Negative Dialectical Feminist Perspective." In *Cultures and/or Globalization*, edited by Barrie Axford and Richard Huggins, 84–110. Newcastle upon Tyne: Cambridge Scholars Press.

Cochrane, Regina. 2014. "Climate Change, *Buen Vivir*, and the Dialectic of Enlightenment: Toward a Feminist Critical Philosophy of Climate Justice." *Hypatia* 29, no. 3 (Summer): 576–598. https://doi.org/10.1111/hypa.12099.

Cochrane, Regina. 2007. "Rural Poverty and Impoverished Theory: Cultural Populism, Ecofeminism, and Global Justice." *The Journal of Peasant Studies* 34, no. 2 (April): 167–206. https://doi.org/10.1080/03066150701516674.

Cuomo, Chris J. 2011. "Climate Change, Vulnerability, and Responsibility." *Hypatia* 26, no. 4 (Fall): 690–714. https://doi.org/10.1111/j.1527-2001.2011.01220.x.

Donovan, Josephine. 2007. "Attention to Suffering: Sympathy as a Basis for Ethical Treatment of Animals." In *The Feminist Care Tradition in Animal Ethics*, edited by Josephine Donovan and Carol J. Adams, 174–197. New York and Chichester, West Sussex: Columbia University Press.

Donovan, Josephine. 2012. *Feminist Theory: The Intellectual Traditions*, 4th ed. New York and London: Continuum.

Echols, Alice. 1987. *Daring to be Bad: Radical Feminism in America, 1967–1975*. Minneapolis: University of Minnesota Press.

Evans, Judith. 1995. *Feminist Theory Today: An Introduction to Second-Wave Feminism*. London, Thousand Oaks, CA, and New Delhi: Sage.

Friedman, Marilyn. 1993. "Beyond Caring: The De-Moralization of Gender." In *An Ethic of Care: Feminist and Interdisciplinary Perspectives*, edited by Mary Jeanne Larrabee, 258–273. New York and London: Routledge.

Gaard, Greta. 2015. "Ecofeminism and Climate Change." *Women Studies International Forum* 49 (March-April): 20–33. http://dx.doi.org/10.1016/j.wsif.2015.02.004.

Garvey, James. 2008. *The Ethics of Climate Change: Right and Wrong in a Warming World*. New York and London: Continuum.

Geus, Marius de. 2003. *The End of Over-consumption: Towards a Lifestyle of Moderation and Self-restraint*. Utrecht: International Books.

Geus, Marius de. 2009. "Sustainable Hedonism: The Pleasures of Living within Environmental Limits." In *The Politics and Pleasures of Consuming Differently*, edited by Kate Soper, Martin Ryle, and Lyn Thomas, 113–129. London and New York: Palgrave Macmillan.

Gilligan, Carol. 1982. *In a Different Voice: Psychological Theory and Women's Development*. Cambridge, MA and London: Harvard University Press.

Gordon, Linda. 2016. "'Intersectionality,' Socialist Feminism and Contemporary Activism: Musings by a Second-Wave Socialist Feminist." *Gender & History* 28, no. 2 (August): 340–357. https://doi.org/10.1111/1468-0424.12211.

Harris, Paul G. 2016. *Global Ethics and Climate Change,* 2nd ed. Edinburgh: Edinburgh University Press.

Hoagland, Sarah Lucia. 1991. "Some Thoughts about 'Caring'." In *Feminist Ethics*, edited by Claudia Card, 246–263. Lawrence, KS: University Press of Kansas.

Jaggar, Alison M. 1988. Feminist Politics and Human Nature. Totowa, NJ: Rowman and Littlefield.

Jaggar, Alison M. 1991. "Feminist Ethics: Projects, Problems, Prospects." In *Feminist Ethics*, edited by Claudia Card, 78–104. Lawrence, KS: University Press of Kansas.

Littler, Jo. 2009. "Gendering Anti-Consumerism: Alternative Genealogies, Consumer Whores and the Role of *Ressentiment*." In *The Politics and Pleasures of Consuming Differently*, edited by Kate Soper, Martin Ryle, and Lyn Thomas, 171–187. London and New York: Palgrave Macmillan.

Luke, Timothy W. 2008. "Climatologies as Social Critique: The Social Construction/Creation of Global Warming, Global Dimming, and Global Cooling." In *Political Theory and Global Climate Change*, edited by Steve Vanderheiden, 121–152. Cambridge, MA and London: The MIT Press.

MacGregor, Sherilyn. 2014. "Only Resist: Ecological Citizenship and the Post-Politics of Climate Change." *Hypatia* 29, no. 3 (Summer): 617–633. https://doi.org/10.1111/hypa.12065.

Noddings, Nel. 1984. *Caring: A Feminine Approach to Ethics and Moral Education*. Berkeley, Los Angeles, and London: University of California Press.

Puka, Bill. 1993. "The Liberation of Caring: Another Voice for Gilligan's 'Different Voice'." In *An Ethic of Care: Feminist and Interdisciplinary Perspectives*, edited by Mary Jeanne Larrabee, 215–239. New York and London: Routledge.

Resurrección, Bernadette P. 2013. "Persistent Women and Environmental Linkages in Climate Change and Sustainable Development Agendas." *Women's Studies International Forum* 40 (September-October): 33–43. https://doi.org/10.1016/j.wsif.2013.03.011.

Ruddick, Sara. 1999. "Maternal Thinking as a Feminist Standpoint." In *Feminist Philosophies: Problems, Theories, and Applications,* 2nd ed., edited by Janet A. Kourany, James P. Sterba, and Rosemarie Tong. Upper Saddle River, NJ: Prentice Hall.

Sassatelli, Roberta. 2007. *Consumer Culture: History, Theory and Politics*. Los Angeles, London, New Delhi, and Singapore: Sage Publications.

Soper, Kate. 2008. "'Alternative Hedonism' and the Citizen-Consumer." In *Citizenship and Consumption*, edited by Kate Soper and Frank Trentmann. London and New York: Palgrave Macmillan.

Soper, Kate. 2009a. "Beyond Consumerism: Reflections on Gender Politics, Pleasure and Sustainable Consumption." *Women, Gender and Research* 18, no. 3–4 (October): 92–100. https://doi.org/10.7146/kkf.v0i3-4.27977.

Soper, Kate. 2009b. "Introduction: The Mainstreaming of Counter-Consumerist Concern." In *The Politics and Pleasures of Consuming Differently*, edited by Kate Soper, Martin Ryle, and Lyn Thomas, 1–21. London and New York: Palgrave Macmillan.

Soper, Kate. 2017. *A New Hedonism: A Post-Consumerist Vision*. The Next System Project. https://thenextsystem.org/learn/stories/new-hedonism-post-consumerism-vision, accessed 28 May 2019.

Soper, Kate. 1990. *Troubled Pleasures: Writings on Politics, Gender and Hedonism*. London and New York: Verso.

Soper, Kate. 1995. *What is Nature? Culture, Politics and the non-Human*. Oxford and Cambridge, MA: Blackwell.

Spitzner, Meike. 2009. "How Global Warming is Gendered: A View from the EU." In *Eco-Sufficiency and Global Justice: Women Write Political Ecology*, edited by Ariel Salleh, 218–229. London and New York: Pluto Press and North Melbourne: Spinifex Press.

Tong, Rosemarie and Nancy Williams. 2019. "Feminist Ethics." In *The Stanford Encyclopedia of Philosophy* (Spring edition), edited by Edward N. Zalta. https://plato.stanford.edu/archives/spr2019/entries/feminism-ethics/, accessed 28 May 2019.

Tong, Rosemarie and Tina Fernandes Botts. 2017. *Feminist Thought: A More Comprehensive Introduction,* 5th ed. New York: Westview Press.

Tronto, Joan C. 1993. "Beyond Gender Difference to a Theory of Care." in *An Ethic of Care: Feminist and Interdisciplinary Perspectives*, edited by Mary Jeanne Larrabee, 240–257. New York and London: Routledge.

Watts, Jonathan. 2018. "We Have 12 Years to Limit Climate Change Catastrophe, Warns UN." *The Guardian*, October 8. https://www.theguardian.com/environment/2018/oct/08/global-warming-must-not-exceed-15c-warns-landmark-un-report.

Zamora, Daniel. 2017. "The Case Against a Basic Income." Translated by Jeff Bate-Boerop. *Jacobin*, December 28. https://www.jacobinmag.com/2017/12/universal-basic-income-inequality-work.

11

PRAGMATIST ETHICS AND CLIMATE CHANGE

Steven Fesmire

Soon after his 90th birthday, John Dewey (1859–1952) was feted at his alma mater, the University of Vermont. Too tired to rise and speak to the crowd in Burlington, he simply said: "I'm thankful for the privilege of living on this good planet, Earth. But living on this Earth has become the supreme challenge to mankind's intelligence" (May 25, 1975? [22283]: Herbert W. Schneider to American Humanist Association).[1]

A century ago in "The Need for a Recovery of Philosophy," Dewey famously advocated a spirit of public engagement by intellectuals: "Philosophy recovers itself when it ceases to be a device for dealing with the problems of philosophers and becomes a method, cultivated by philosophers, for dealing with the problems" of humanity (1917, MW 10:46).[2] By facing widely shared problems and helping to guide inquiry into them with fresh hypotheses, Dewey argued that philosophers could recover a coherent social function (cf. Fesmire 2015). Those who join Dewey in seeing philosophy's public function as cultural interpretation and criticism are not likely to see professional philosophy's recent past as the best guide to its future, isolated as it has been from contemporary conflicts, disparities, divisions, and drift. Nevertheless, philosophers in areas such as bioethics and environmental ethics have for decades been charting a course to determine, in Dewey's words, "the character of changes that are going on and to give them in the affairs that concern us most some measure of intelligent direction" (1930, LW 5:271).

Building especially on Dewey, this chapter explores some features of pragmatic pluralism as an ethical perspective on climate change. It is inspired in part by Light's work on climate diplomacy (e.g., 2013; 2017) in 2013–2016 as Senior Advisor and India Counselor to the U.S. Special Envoy on Climate Change in the Obama administration, and by Norton's environmental pragmatism (e.g., 2005; 2015), while drawing more explicitly from classical pragmatist sources than either

Light or Norton. My primary aim in this chapter is to characterize, differentiate, and advance a general pragmatist *approach* to climate ethics, programmatic though it is, rather than to defend specific governing principles, formulate or prescribe climate policy recommendations, or advance action items.

In the main, contemporary climate ethicists focus on assessing policy options and making prescriptions. This is unsurprising, as much of the agenda for climate ethics has been set by meetings responsive to the UN Framework Convention on Climate Change. However, an exclusive focus on policy assessments would take the "practical" too narrowly. Despite the practical bearings of debating policy options, this chapter does not, for instance, weigh in on controversies regarding Stern vs. Nordhaus on ethically appropriate ways to discount future costs and benefits of climate action, or the relative prioritization of adaptation and abatement, or how best to assess expected future value in lieu of an excessively moralistic precautionary principle. My main line of argument is that we are suffering culturally from a sort of moral jetlag due in part to moral fundamentalist habits, and that a pragmatic turn—in moral theory generally and climate ethics particularly—would be salutary for our recovery if philosophers are to speak more effectively to "wicked problems" on a warming planet in a way that aids public deliberation and social learning.

A Moral Fundamentalist Pledge of Allegiance

I sometimes ask my students to bring examples of people weighing in on a contemporary issue such as climate change, immigration, race, gender, reproduction, or marriage. How many of those people, I ask, would knowingly raise their hand and pledge the following?[3]

> *There's a single basis of moral and political life, and this supreme basis determines the right way to proceed. I have access to this supreme basis. When others don't agree with me, it's because they have the wrong faith commitments or they aren't analyzing things properly. Agreement with me is a prerequisite to solving our problems. Consequently, I have nothing to learn about these matters from those who disagree with me. Their participation is at best an irrelevant distraction and at worst an evil to be defeated. My diagnosis of the issue has precisely captured all that is morally or politically relevant. It's exhaustive, hence beyond revision and reformulation.*
>
> *Fesmire 2017*

After my students and I swap stories about those who might blithely take such a pledge, we invariably conclude that the pledgers are outnumbered by their counterparts: conservatives, liberals, and radicals who would, upon conscious reflection, reject this outright as cocksure arrogance. Moreover, it quickly becomes clear that this pledge—especially beginning with the third sentence—does not speak to the sort of people my students wish to become.

And yet, our class conversation continues, how many of us certified broader-minded souls act as though complex problems come prepackaged with our singular interpretation or framing of them? Do we prejudge and offhandedly dismiss alternative diagnoses of shared moral issues? However open-minded we may seem to ourselves, do we react to others as though we are navigating with the one, universal moral compass? Are our real behaviors implicated in this pledge most of us would consciously disavow?

At this point, shifting uncomfortably in our complicity, my students and I pause to explore our cognitive dissonance. Perhaps we are merely hypocrites, parading open-mindedness while betraying its opposite. Or perhaps we are beset with a neural vestige of moral tribalism, which some think might finally be enlightened by a universal morality (Greene 2013). But I think there is also something more philosophically interesting at work. To explain, I introduce in this section the terms *moral jetlag* and *moral fundamentalism*, then single out two tell-tale features of *wicked problems*, a vague and overused notion that nevertheless has some practical traction for climate ethics.

Dewey proposed soon after the atomic bombings of Hiroshima and Nagasaki that citizens of techno-industrial nations suffer from "cultural lag" (1945, LW 15:199–200; cf. 1929, LW 4:203–28). I will call this "moral jetlag," a condition in which most of the basic alternatives we have on hand to think and talk about moral and political life, from customary moralizing to sophisticated theorizing, were developed, canned, and pickled on a shelf so long ago that they now lag far behind the multi-faceted problems that our values must speak to. Our moral imaginations are nourished in this conflicted social matrix, resulting in moral jetlag.

In *Reason in a Dark Time*, Jamieson explores an implication for climate change (2014, ch. 5) of what I am calling moral jetlag. At least in the Anglophone world, the commonsense prototype of a harmful activity—one for which we ought to feel and be held responsible—is one that has negative consequences that are immediate, localized, intentional, and directed toward individuals. But this conception of responsibility for harm is eerily out of step with the actual conditions of contemporary lives in complex systems. For example, the greatest harm caused by local greenhouse gas emissions is long-term, widely distributed, unintentional, and not directed toward individuals. Partly on this basis, Jamieson concludes that climate change presents challenges that "go beyond the resources of commonsense morality" (2014, 6). In this context, moral jetlag is characterized by inherited moral concepts and theoretical frameworks that are too narrow, homogeneous, and individualistic to adequately meet many of the problems of techno-industrial civilizations, exemplified here by a lack of fit with anthropogenic climate disruption.

Due in part to this moral jetlag, many people have an uneasy relationship with their inherited "moral fundamentalism," which can be defined ostensively as the cluster of habits exhibited in "the pledge" activity. The term moral fundamentalism was coined by Johnson (2014) as a synonym for extreme moral absolutism.

Its rejection as a general outlook has entered politics and policy in part through extensive research on "wicked problems," inspired by Rittell and Webber (1973) and now ubiquitous in the transdisciplinary field of environmental studies, along with research on democratic decision-making by pragmatists such as Norton (2015; cf. Sarkar and Minteer 2018) and proponents of deliberative democracy such as Niemeyer (2013).

Moral fundamentalism is a vice because it obstructs communication, constricts our deliberative excursions *ex ante* into what is possible (cf. Fesmire 2003), and underwrites bad decisions. It makes the worst of our native impulses toward social bonding and antagonism. And when we oppose others' moral fundamentalism with our own, we drive our society's us–them wedge even deeper. Exercising and reinforcing moral fundamentalist habits gives them safe haven and perpetuates the root problems. Meanwhile, reactionary nihilism and extreme moral skepticism are merely moral fundamentalism's mirror image, setting up a stock false dilemma between extreme relativism and absolutism that is yet another symptom of our moral jetlag.

Research on wicked problems suggests a more precise definition of moral fundamentalism, one that offers a resource for climate ethics. Without canvassing the many senses of "wickedness" in the policy literature, at least two necessary features can be identified that cut through the noise: when we say a problem is wicked rather than benign, we hypothesize at least that (1) there is no single definitive solution and (2) the way we formulate a problem, and the way we appraise success in dealing with it, are themselves at issue. When confronting wicked problems, as Norton observes, "it is necessary to problematize problem formulation itself" (2015, 37), because in these cases even the most sincere and informed participants formulate problems and interpret facts differently. Many contemporary problems are candidates for wickedness in this sense, especially in complex systems: for example, climate, ecosystems, international relations, economic systems, food systems, legal systems, governmental institutions, inter-governmental institutions, and educational institutions. In climate ethics, for example, perhaps there just is no theoretically correct balance between a focus on vulnerable individuals (i.e., climate justice) and a focus on systems.

Observing that many problems have similarly intractable tensions, Gardiner (2017) argues that wickedness is an unnecessarily vague concept. Nevertheless, the concept has proven useful in policy studies for highlighting that moral and sociopolitical life-as-usual is messier than many theorists have taken it to be (cf. Thompson and Whyte 2012).

Married to this literature on wicked problems, a moral fundamentalist might be redefined as someone who holds that there is (1) a single right way to diagnose moral or political problems and (2) a single approvable practical solution to any particular problem. This definition clarifies the sense in which moral fundamentalism is a vice. When we see a moral or political problem only as given, not taken, the chief problem is presumed to be that others do not get the problem (Norton

2015). Or the main problem is presumed to be the general failure of others to bow to our brilliant solutions. Never mind the unnoticed parts of the mess occluded by our well-defended general principles, which are often presented as being value-neutral and free of interest-driven rationalizations and inherited biases. When we suppose our diagnosis of the problem is exhaustive, we autocratically predefine what is relevant and we prejudge alternative formulations without dialogue.

What happens, then, to opportunities for democratically learning our way toward a more sustainable future across a spectrum of values, beliefs, and concerns? It is a truism that people happily weigh in on matters concerning which they are incompetent, but this retort simply clarifies democracy's uphill struggle. No problem is so bad that we cannot make it worse through our way of dealing with it. In public disputes, competing moral fundamentalist camps restrict the sympathies of in-groups to an exclusive channel. This channel may be—and often is—*progressive* in one dimension of a problem, but typically at the cost of being *regressive* with respect to marginalized concerns. Moral fundamentalist camps demand such single vision to the logical exclusion of attempts to secure shared toeholds to debate and achieve controverted *social goals* (goals we can only achieve together) like security, health, sustainability, and justice. For example, Thompson observes of the food movement that "Advocates of both biotechnology *and* organic systems too often compare the most advanced and optimistic interpretation of their favored approach to the least successful applications of the alternative" (2015, 252). The result is dichotomized either/or thinking, attended by puritanical tendencies to ignore context, oversimplify, and dogmatize (Boisvert and Heldke 2016). In this way, wherever moral fundamentalist habits find a safe haven, they encourage parochial antagonism toward excluded standpoints, closure to being surprised by the complexity of many situations and systems, neglect of the context in which decisions are made, obtuseness about one's own truncated framework, and a related general indifference to public processes and adaptive policies.

A Wicked Case for Pragmatic Pluralism in Moral Theory

Moral theory gets its social value by enlarging perceptions and making us more sensitive to the world. Moral fundamentalism, along with its reactionary sidekick nihilism, is among our chief obstacles to cultivating such sensitivity. To help us navigate contemporary entanglements, theorizing should avoid legitimizing and perpetuating these obstacles. Yet moral and political theories too often prop up moral fundamentalism by unintentionally legitimizing its one-way feature, as a by-product of monistic appeal to a supreme moral principle (e.g., Gewirth 1978), value, standard, law, concept, or ideal that purportedly exhausts whatever is morally relevant. Of course monistic theorists are rarely themselves moral fundamentalists, but they inadvertently reinforce and legitimize moral fundamentalism's hidden major premise: the dogma that there is a single conceptual home range of moral action.

Moral fundamentalism cannot logically stand without its monistic premise. Or to approach it the other way, only a monist can be a logically consistent moral fundamentalist. Pluralists (Rossians, pragmatists, cosmopolitans, deliberative democrats, etc.) are moral fundamentalists at the cost of logical coherence. Of course the fact that monism and moral fundamentalism rely on a shared premise does not on its own refute that premise. But it suggests that setting the monistic premise aside, at least methodologically, could place practical ethics on a stronger footing that better checks and compensates for our complicity in moral fundamentalist behaviors, regardless of which moral philosophy captures our imaginations. We especially need these checks when dealing with climate disruption and other anthropogenic drivers of rapid global change (e.g., habitat depletion, invasive species, nonpoint source pollution, and antibiotic resistance), in which problem formulation across disparate groups is itself among the key problems.

Monists in moral theory abstract some factor of moral experience as central and uppermost, hypostatize it, then treat it as the self-sufficient starting point for moral inquiry and the bedrock for justification. The popular habit of singling out one trump value among a wide range of relevant values tracks the same pattern, as when economic criteria are presumed by mainstream environmental economists to have supremacy over other key values (aesthetic, recreational, ecological, etc.) (Norton 2005). The simpler the problem, the more likely a monistic reduction is to work. But contemporary moral and political conflicts are rarely so simple that a correct rational analysis could, even in principle, sweep the path clear toward what is "truly" good, right, just, or virtuous (cf. Taylor 1982; Williams 1985).

As Dewey observed, under the restrictive monistic assumption legitimized by traditional moral and political theorizing, conflict and diversity are merely apparent (1930, LW 5:279–288). A situation may *seem* to be a quagmire, the supposition runs, but closer examination, or more data to feed into our utility calculations (Singer 2015), will in principle always reveal that there had been a single right, fair, or best path through the territory all along. If there is a unitary conceptual home range of moral or political action, the moral conflict boils down to mere hesitancy on our part about what to choose. What is good or virtuous or right is presumed to be already licit, ready to be laid bare by intellectual analysis.

Dewey proposed, however, that traditional key moral categories such as good, duty, and virtue express different experiential origins. He argued that none operates as the bottom line that can accommodate all that is of moral worth in the rest. If one concept is neither logically derivable from another nor translatable without remainder into the terms of another, then, he contended, distinctive experiential phenomena in moral life cannot be blanketed by a single covering concept (see Fesmire 2019).

Dewey consequently saw little place for zero-sum disputes in theory or policy assessment (cf. Edenhofer and Kowarsch 2015). Historical ethical theories and traditional codes of conduct are resources for inquiry, not finalities to be accepted or rejected wholesale. For example, Broome's utilitarian notion in *Climate Matters* that "the good of the world is the arithmetic total of people's well-being" allows him to develop a project with a distinctive set of dominant emphases, angles, and inferences. This can contribute to democratic discourse. But from Dewey's pragmatist perspective it cannot finally do something that Broome apparently wants to do, which is to help determine in advance whether to adopt a policy by taking "a sort of weighted average across the portfolio of all the possible amounts of well-being that might result from our policy" (Broome 2012, 116). The utilitarian aggregator's inference flows not from logic but from a hidden premise of theoretical correctness, which presumes that an account of metaethics and normative morality can be given straightforwardly in terms of one supreme root.

Akin to Dewey's pluralistic account of the tangled terrain of moral action, Latour (1993) argues that "imbroglios" typify moral experience: that is, moral predicaments are entanglements of often-incompatible forces. From a pragmatist perspective, this relative incommensurability of forces presents a *practical* problem (not primarily a theoretical one): if diverse situational factors are already in tension with each other so that we are tugged in multiple ways, then one-way decision-making leads to overly simplistic normative prescriptions that ignore or relegate factors relevant to intelligent choices. This is analogous to the logical fallacy of causal reductionism, assuming a single cause for a complex outcome and ignoring multiple conjoint variables.

If moral action is heterogeneous in its origins and operations, and typified by underlying tangles between irreducible forces, then ethical monism's usefulness to moral understanding is limited. That is fine. Indeed, as Appiah (2017) argues, the articulation of one-sided idealizations is a personal or collective help in specific contexts, and the pluralist for her part is not immune from obtuseness. The problem with traditional monism, however, is that the quest for a plumb line of reason that will square our moral lives to the world and impose order on deliberation is (1) philosophically dubious, and (2) fastens the linchpin of moral fundamentalism.

Again, we should delegitimize moral fundamentalism if we are to recover from our moral jetlag and inquire more effectively into wicked problems. More specifically, far from being an antidote to what Callicott wittily dubs moral pluralism's analogue to "multiple personality disorder" (1999, 175), moral monism's greatest risk is that it will obstruct or exclude inquiry into situational tensions and concerns that are off-the-radar of our idealizations.

There is nothing anti-theoretical in these observations. I am raising a question about how to theorize more effectively, in a way that helps to create a context for making better decisions together. If being theoretically correct in ethics implies, as a regulative ideal, a completely enlightened standpoint secured in advance of confronting difficulties in particular contexts—a standpoint from which our general habits of moral thinking will, with tweaks here and there, be adequate to meeting every relevant situation—then the quest for it increases our moral jetlag.[4]

Traditional monistic theories are helps. But their good work becomes the enemy of better work when we fail to remember that whatever we see with their help is always situated within what we do not see; what they put us in touch with is situated within what is inconspicuous and ungrasped (cf. 1925/1929, LW 1:44). As James said, "Something always escapes" (1977, 145). When we forget this, we pay for conceptual clarities by forgetting relevant aspects of the troubled existential subject matter that occasioned our inquiries in the first place.

In sum, traditional moral theorizing's "zeal for a unitary view" (1930, LW 5:288) tends to oversimplify and standardize moral life in a way that is ill-suited to navigating complex, indeterminate systems. This is a problem for climate ethics, and it arises in part from the monistic one-way feature that traditional moral theory shares with moral fundamentalism. The resulting moral jetlag is troubling, but not merely because we persist in using benighted ideas. Moral jetlag is worrisome because of the lack of fit between simplistic habituated outlooks and real, unsettled, on-the-ground circumstances.

To ameliorate the morasses we face, we need the moral clarity (see Neiman 2008) of theoretical idealizations, not merely new iterations of postmodern suspicions of them. In the idiom of Sherlock Holmes, such idealizations keep us concentrated on what is vital instead of being dissipated by what is incidental. But in place of incorrigibility and oversimplification, we should reintegrate abstract ethical theorizing with the entanglements of direct experiences.

Pragmatism and the Ends-Means Continuum

Like other ethical naturalists, pragmatists argue that we can intelligently deal with problems and direct ourselves toward desirable goals, both individually and collectively, without transcendental standards or *a priori* deductions that hide from inspection even as they pretend to guarantee the validity of judgments. More distinctively and controversially, though in common with many other strong pluralists, pragmatists tend to regard the mainstream quest in moral theory for the central and basic source of normative justification as outdated. For the pragmatist, the moral theorist's job, in Dewey's representative view, is to systematically work through and generalize about situations in which the way forward is not well lit, when multiple paths beckon, and when incompatible goods and colliding duties "get in each other's way" (1932, LW 7:165).

The term "pragmatist" has done as much to muddle as to clarify. In 1908, Lovejoy identified thirteen conflicting senses of pragmatism then in vogue among philosophers (Lovejoy 1963). "Pragmatism" in colloquial English is a bit more definite, with arguably two principal meanings. It primarily suggests the tempering of ideologies with practicality, the balancing of principles with achievable outcomes, or simply flexibility amid contingencies. To call someone a pragmatist popularly suggests a counterweight to compensate for pie-in-the-sky ideals or dogmatic ideology. Worldly pragmatists ensure that some portion of our ideals may be realized, which they presume is better than none at all. The word secondarily means, again outside of academic philosophy, pursuit of the most expedient means to satisfy a self-interested desire, often associated with an outlook in which political actors take short-sighted gambles that may well backfire (e.g., Jotzo 2016).

If we shave off its anti-intellectual and anti-theoretical connotations, the first popular sense above may capture something of the "Yankee pragmatism" in Dewey's theorizing. But this requires a string of qualifications: Dewey's corpus rings with criticisms of shallow practicality, unprincipled realpolitik, machinations toward fixed ends, atomistic individualism, acquisitiveness, and American swagger. His philosophical reconstructions cut much deeper than the pragmatism of common parlance. Indeed he wrote in a 1940 letter: "The word 'pragmatism' I have used very little, and then with reserves" (September 6, 1940 [13667]: Dewey to Corliss Lamont).

Dewey expanded and rigorously systematized Peirce's and James's pragmatisms as a means for reconstructing philosophy to meet evolving difficulties. Dewey's pragmatism is, minimally, the critical attempt to replace inherited beliefs and distorting prejudices with intelligent inquiry. Had he ever formulated a pragmatic maxim to clarify just what he thought made inquiry more *intelligent*—which was one with specifying what makes it *experimental*—it might spotlight his emphasis on the ends-means continuum: Always state your ends in terms of the means you plan to use to achieve them. Dewey analyzed the feedback loop that had always been implicit in the familiar dichotomy between what people mean to do, on the one hand, and what they've actually done, on the other hand. For example, if you are ultimately aiming for climate justice, equitable distribution of emissions, the right to sustainable development (Moellendorf 2014), or the rights of future generations (Gardiner 2011), then state these ends in terms of what you intend to do or advocate to bring them about. Then do your best to confer and pool experiences so that you track *all* of the rippling consequences of those means and not just the ones that suit your agenda. Review—through colloquy, not soliloquy—what you have actually done, and revise what you mean to do next accordingly.

Alternatively, framed negatively as a cautionary pragmatic maxim: Beware anyone's ends which are asserted *ipse dixit* or autocratically as finalities or absolutes rather than "in terms of the social means" being proposed (Mead 1930, 104–105). Take security as an end. Ronen Bergman's book *Rise and Kill First* (2018a) shows that Israel's longstanding practice of targeted assassination—drone strikes, bombings, shootings, and poisonings—has resulted in many "tactical successes" that have dramatically worsened diplomatic relations. Yet the shared social goals of conflicting actors in this region can only be achieved through diplomacy, not imposed through the method of force. Bergman observes in an interview:

> They felt that at the tip of their fingers, they can hit someone way beyond enemy lines, deep in the enemy state and solve the problem, and therefore,

> they do not need to turn to statesmanship or political reconciliation. And therefore I think the story of the use of these special means is a series of extraordinary tactical successes but, at the same time, a disastrous political failure.
>
> *Bergman 2018b*

When we track only the consequences (the "ends" in Dewey's idiom) that we *meant* to bring about (such as eliminating an enemy), we do not usually feel responsible for the collateral side effects of our choices. The only defensible alternative to singling out and favoring some fragmentary preferred consequence as "the" end is to evaluate desires, ends, and consequences as themselves "means of further consequences" (1938, LW 13:229).

The mad, anti-pragmatist maxim "the end justifies the means" arises from the warped idea that ends are to be valued irrespective of means. We can imagine a fictional universe in which this is not absurd: by way of advice to science fiction writers, it would have to be a universe in which miracles intervene to keep the means we employ from having their usual side effects. It would also have to be a universe in which things reliably turn out fine even when moral agents crave particular ends so much that they fail to forecast unintended consequences cascading through the entire network of relationships and events in which their lives are embedded, "no matter how intrinsically obnoxious" these consequences are (1938, LW 13:228). To offer an illustration on the lighter side: "Where are you going?" Hobbes asks. Calvin, walking with a bucket in hand, replies: "I'm going to the other side of the lake."

"What's the bucket for?"

"I'm going to drain the lake" (Watterson 1988).

Returning to *our* universe, perhaps a suitably broad, multi-scalar, and long-range conception of ends could indeed justify the means. I will leave that Pandora's box closed except to caution that Dewey's ethics does not fit most familiar forms of consequentialism (cf. Pappas 2008). For instance, Dewey joined Kantians in criticizing utilitarians for overlooking the practical bearings of attitudes and predispositions, of will (1932, LW 7, ch. 12). Nevertheless, he thought that his naturalistic deconstruction of "the end justifies the means" was far more plausible than Kant's anti-naturalistic and anti-pragmatic notion in *The Critique of Practical Reason* that "the concept of good and evil must not be determined before the moral law ..., but only after it and by means of it" (Kant 2002, 37; cf. Rawls 1971, 31). The pervasive notion in deontological ethics that some objectives are ends-in-themselves is, from Dewey's perspective, incoherent and dangerously short-sighted about the way events relate to one another over time and scale. There is indeed something more basic to honesty than that it is the best policy, and something more to kindness than that it is reciprocated, but this does not entail that honesty and kindness are coherently defended as ends-in-themselves.

That we can rarely if ever do a single thing was formulated by Garrett Hardin as the first principle of ecology. For example, when I set out to make a cup of coffee,

I may inadvertently affect the population of migratory songbirds whose nesting grounds in Central America have been bulldozed and burned to plant coffee plantations. The fact that choices are pregnant with unanticipated connections, both proximal and distant, has more recently become par for the course in the field of ecological economics. Some decades earlier than Hardin, in *Theory of Valuation* (1939), Dewey explored moral implications of such commonplaces from the physical sciences: "Nothing happens which is final in the sense that it is not part of an ongoing stream of events" (TV, 1938, LW 13:229). We usually do more than we mean to do. Whatever end we bring about will engender other existential connections and possibilities, so it must be evaluated empirically as a potential help or hindrance.

Kyoto Hedgehogs and Parisian Foxes

"The Hedgehog and the Fox" (1953) is Berlin's famous riff on Archilochus's saying that "The fox knows many things, but the hedgehog knows one big thing." There are two kinds of intellectuals, Berlin proposes, tongue slightly in cheek: monistic hedgehogs and pluralistic foxes. He contrasts the "centripetal" actions and ideas of the hedgehog with the "centrifugal" ones of the protean fox.

How do we make better decisions? To pursue a rhetorical simplification in Berlin's spirit (inspired by Norton 2015), with a binary heuristic that would amount to caricature if applied *in toto* to any individual moral or political theorist, the monistic hedgehog—with its one big idea—looks for moral answers and progress in the wrong place. We might imagine a physician, Dewey prodded, who seeks to heal patients in light of some ultimate, final, complete, and universal ideal of perfect health, when the situation calls us to aid living processes of recovery (1922, MW 14:196). In the face of dynamic complications, whether in medicine or climate policy, this would be quackery. In contrast, the pluralistic fox approaches matters *in medias res*, starting with the entanglement. The hedgehog is a detached spectator, while the fox emphasizes our role in actively remaking situations through reflective choices and deeds. Instead of disengaging as a calculator, hovering jurist, or legislator, the fox is an active, imaginative, and experimental participant.

Do we serve our students well if we merely educate them to judiciously weigh matters so that the balance tips toward a purportedly optimal policy supported by general principles derived prior to inclusive participation in real situations? Economizing deliberation with a pre-determined, rule-governed metric can be helpful for many purposes. But no matter the amount of data plugged in or how well argued our principle of justice, insofar as an approach fails to also prioritize sensitivity to context, creative social inquiry, and experimental understanding of complex underlying structures, its end result is too often reminiscent of an off-handed criticism that Dewey once made about "popcorn" solutions: put the right amount in the right mechanism and you get some "unnutritious readymade stuff" that will not sustain anyone for long (1951.02.14 [14090]: Dewey to Max C. Otto).

Exemplifying a foxlike approach to climate ethics and politics, the pragmatist starts with the concrete situation, not an abstract concept such as equity. Take the responsibilities of countries for reducing greenhouse gas emissions. Light highlighted in the buildup to the 2015 Paris climate agreement "a problem for any treaty that includes a notion of equity that conceives of the atmosphere as a global commons in which individuals, groups, or nations may claim shares." Many countries had been following India in advocating for equity in *historical* per capita emissions as the best way to determine a country's fair share of allowable emissions. Light observes that this abstractly defensible allocation places the United States in a severe "carbon debt" (2013, 33). Moreover, there is no clear path in federal law to restrict its emissions to accord with such an allocation. This presents a domestic hurdle for the United States to join an ambitious agreement, yet American participation is necessary to achieve "some modicum of climate safety" (30).

In *One World*, Singer acknowledges that the historical version of the principle of per capita equity is an internal political hurdle for industrialized countries (2004, 195). As a political compromise to bring about the best outcome, Singer proposes a forward-looking principle of "equal per capita future entitlements to a share of the capacity of the atmospheric sink" (194). An emissions trading scheme is compatible with this principle, and Singer thinks such an approach is politically plausible, though it may eventually require United Nations sanctions against the United States analogous to those against apartheid South Africa (198).

The difference between Singer and Light does not pivot on a disagreement about the "optimal allocation of global reductions in emissions in the abstract," nor on how this should be carried forward in a global climate treaty that will hold everyone's feet to the fire (Light 2013, 35). Light does not deny that there may be a theoretically defensible optimal emissions allocation for countries, or more-or-less optimal ways to deal with high short-range costs that have uncertain long-range benefits (see Broome 2012, ch. 8), but he insists that we remain clear about the aim of such debate: to sort out what to do, not to agree in advance on justifications or on the single correct way to reason about the matter. So he focuses instead on a decision process to elicit the generative possibilities of an international situation that had been shackled by an overly legalistic approach that was insensitive to intractable tensions.

This may be clarified by contrasting the Kyoto Protocol with the Paris Agreement. At risk of oversimplification, the Kyoto Protocol took a top-down, punitive approach tethered to an abstract principle of equitable emissions. According to Light, by emphasizing *ends fixed in advance*, with penalties attached, Kyoto precipitated a race to the bottom when it came to agreeing on binding targets.[5] The Paris Agreement took a bottom-up, "pledge and review" approach (aka "shame and blame"), which Light regards as exemplifying a pragmatic emphasis on adaptive action, making it more likely that evolving situations will be met creatively and with higher ambition. While retrospectively acknowledging the unreliability of the United States as an international actor in non-binding agreements, Light argues that relying on the benefits and backlash of being judged a good or bad international actor—especially as the stakes of climate disruption continue to rise—"will not guarantee success for achieving some level of climate stability, but it will create an environment in which that will be more likely" (2017, 495). Additionally, Light argues, this approach makes it more likely that climate justice issues will be effectively addressed, such as the necessity for the United States to return to a coalition that agrees on a high-ambition interpretation of shared but differentiated responsibilities—that is, recognition that those most to blame are generally least vulnerable to the harms of climate disruption (2017; cf. Caney 2005).

In Light's view, Kyoto's justifications about why we must deal with the problem were monistic, while Paris's approach was pluralistic and culturally contextual, at least in the sense that a country's pledge to the world has to make sense domestically. Light observes, for example, that India's Prime Minister Modi understands dealing with climate disruption primarily as a religiously bound moral obligation tied up with Hindu scriptures, not primarily as a utilitarian problem about how to secure the aggregate welfare of India's growing population. By encouraging pledges "to be embedded in their cultural contexts," countries under the Paris Agreement need not share views about *why* they should respond to the problem, but they can nevertheless create a context of inquiry in which they may converge on effective policies.

Light adds that Paris is ameliorative and encourages the celebration of intermediate progress as part of five-year plans. This raises a question about what counts as "progress" for a pragmatist, given the absence of any hedgehoggish idea of a final and ultimate good for measuring it. Dewey argued that achievements in our dealings with intrinsically messy problems are real, and they are to be celebrated. But they are not measurable by any rigid "general formula of progress" (1922, MW 14:196). He rejected the two most influential variations of the misguided quest for an absolute standard by which to measure progress: (1) the juvenile notion that progress "means a definite sum of accomplishment which will forever stay done, and which by an exact amount lessens the amount still to be done … on our road to a final stable and unperplexed goal," and (2) the popular though foolishly pessimistic notion that all achievements are negligible in comparison to ultimate and perfect goods (1922, MW 14:197–198). From this angle, whereas Jamieson (2014) is inclined to see the 2009 Copenhagen climate meeting as a disaster from which no good could come, Light sees Copenhagen as a prelude to bringing together a "high-ambition coalition" of odd bedfellows in the Paris Agreement, a coalition that might take things to the next level.

Returning with a grain of salt to Berlin's binary heuristic, the hedgehog, at home in a settled world, asserts *a priori* that its job is to show which antecedently defended, (relatively) static principles should govern choice. So the hedgehog focuses on getting the theory right and impersonally deciding whose values measure up to its supreme metric (whether taken to be constructed or foundational). Start with getting the theory right, and the rest follows! Meanwhile, the fox, at home in an open world and spotlighting the fallibility and incompleteness of any decision or policy, attends to controlled adaptive *processes* through which we may *interpersonally* decide what to *do*, listen, pursue creative leads, take stock, and correct our mistakes. Hedgehog processes (more akin to Kyoto) are expert-governed to predetermine a metric that will yield the right, optimal, or ideal outcome; foxlike processes (more akin to Paris) engage communities of inquirers, thereby fostering growth in the public imagination as both means and end.

We all have hedgehog habits, but it might help to ask them to take a back seat for a while. They do not do much to check and destabilize our inveterate moral fundamentalism. Foxlike adaptive habits have been underdeveloped in our culture(s) despite their compensatory value for dealing with widely shared problems in complex systems. While acknowledging the pragmatic worth of hedgehog habits (Dworkin 2011), there is at least no logical inevitability in their codification remaining the primary and overriding focus of moral and political theorizing about climate change and sustainability.

Pragmatism and Moral Theory

Although Dewey's approach to moral and sociopolitical theory was more radically contextual and problem-centered than many of his twenty-first century allies (see Pappas 2019), pragmatists tend to find common cause with those criticizing the quest for a self-sufficient "ideal theory," as Rawls (1971) called his idealized "original position" approach to a well-ordered society. The Rawlsian in climate ethics tries to discern policy options in which free, equal, and autonomous rational contractors can fully comply with the requirements of justice. Rawls proposed a division of labor between ideal and non-ideal theories. The former's job is to determine "what a perfectly just society would be like" (1971, 8–9), whereas non-ideal theories are tasked with discerning principles to deal with non-ideal conditions in which people do not comply with the principles of justice, as with war or racial oppression, or in which circumstances make perfect justice unrealizable. We need to start by constructing an ideal theory, Rawls thought, if we are to construct a moral compass for dealing with non-ideal conditions. His critics argue that he was mistaken.

Recent critics of "ideal theory" approaches to moral and political theory (an approach also illustrated by Nozick (1974) and arguably Dworkin (2000)) include, to various degrees, Mills (2005, 2017), Anderson (2009, 2013), Pappas (2008, 2019), Sen (2009), Valentini (2012), Gaus (2016), and Appiah (2017). Anderson, for example, has influentially argued that Rawls's approach to an ideal society blinds us to race-based, gender-based, and other social injustices to a degree that is "epistemologically disabling" (2013, 5).[6] These critics propose shifting to a non-ideal starting point for sociopolitical inquiry. They insist that values have to be appraised in light of the particular experiential contexts and purposes that generated them.

Wedges between ideal and non-ideal theories are easily overstated, and they often converge more than is acknowledged in the current debate. For example, Moellendorf's *The Moral Challenge of Dangerous Climate Change* fronts an "Antipoverty Principle" (2014, 22) that is a variation on Rawls's worst-off principle, and in justifying this principle he rejects dichotomizing theoretical justifications from practical applications (4). Nevertheless, Appiah enjoins, even a plausible and practically informed ideal theory "doesn't help much in the circumstances of an actual non-ideal world" (2017, 120).

Dewey approached and evaluated ethical and political theories not on analogy to logical or mathematical problems, but as experiments in "living together in ways in which the life of each of us is at once profitable in the deepest sense of the word, profitable to himself and helpful in the building up of the individuality of others" (1938, LW 13:303). In the contemporary ambit of Kitcher and Johnson, Norton and Light follow Dewey in being focused on moral inquiry as an experimental process rather than merely as incessant verbal argumentation. What Kitcher (2014) calls "the ethical project" is a process in which, as Johnson explains it, we actively try out "various modes of behavior (verbal and nonverbal), various institutional structures, and various life strategies." Moral theorists have neglected the way experiments in living also constitute "arguments" for and against various practices, in the sense that certain practices address or fail to meet problems. Such arguments, Johnson argues, are enactive, embodied, and embedded (2014, 126), and they should be more central to the future of moral theory.

Dewey saw variability in valuing and valuations as a starting point for constructive inquiry, rather than as deviations to be suppressed or intellectually standardized in the name of ethical truth. We would be better off if we would experiment with how far we can go to create a context for ongoing shared inquiry—again, not only verbally arguing, but also "on-the-ground experiments in living" (Johnson 2014, 126; cf. Kitcher and Keller 2017; cf. Mill 1986)—in which we steer between what Elgin aptly calls "the absolute and the arbitrary" (1997).

Our experiments in living together involve ideals and idealizations through which we appraise moral alternatives, as Appiah has argued (2017), but they must proceed without privileged access to an ideal standpoint. Favoring such a non-ideal, context-steeped starting point may seem, especially to deontologists, to endorse a "Missouri Compromise" attitude, bartering away human rights and dignity for the sake of a mealy-mouthed pragmatic solution. But what we judge to be progressive or regressive is ultimately, in Kitcher's words, "something people work out with one another. There are no experts here" (2014, 286). On this view, the main job of the expert in moral theory is not to tell us the right thing to be doing, but to help us make better decisions together. Nevertheless, Kitcher and Johnson argue that rejection of aperspectival ethics "in no way keeps us from making reasonable claims about" the suitability of certain "values, principles, and practices" over others (Johnson 2014, 129; Kitcher 2014, 210ff.). Given a non-ideal starting point, facing non-ideal conditions, Kitcher and Johnson recommend an experimental and democratic pragmatic pluralism as a strategy for moral and political inquiry into what we ought to deem progressive.

Environmental Pragmatism

In their edited volume *Environmental Pragmatism* (1996), Light and Katz define environmental pragmatism as "the open-ended inquiry into specific real-life problems of humanity's relationship with the environment" (2). Franks et al. (2017) are correct, as far as they go, in asserting that environmental pragmatism is "an approach to environmental ethics that emphasizes the need for environmental activists and academics to open-mindedly engage with people's existing environmental attitudes and behaviors if they are to have any influence over them" (Franks et al. 2017, 13). But more should be said (e.g., see Hourdequin 2014, ch. 8).

The standout feature of environmental pragmatism, as represented for example by Norton, Light, McKenna (2018), Thompson (2015), Minteer and Collins (2005), Minteer (2011), Minteer et al. (2018), and Weston (1991), is rejection of the mainstream attempt to find a single defensible paradigm with which we must align ourselves. Specifically, whatever their own eco-ontologies, pragmatist environmental ethicists do not respond to anthropogenic climate disruption by prioritizing a revolutionary attempt to convince doubters that natural systems have intrinsic value. Instead, they tend to focus more than monists on ameliorative processes for resolving disagreements, on making workable, ecologically informed decisions (Minteer and Collins 2005). Fellow pragmatists may or may not additionally concur with Norton's controverted "convergence hypothesis" that broad-scope anthropocentric arguments usually justify the same policies as ecocentric arguments (1991).

This does not imply that environmental pragmatists sidestep the evaluation of principles in climate ethics and policy, such as tensions between the precautionary principle and cost-benefit analyses (see Gardiner 2011; Broome 2012, ch. 7) or debates underlying the discount rate (Broome 2012, chs. 6–8). Ignoring principles would, as Broome argues, undemocratically leave the consequences of following divergent principles solely to those with technical expertise and would fail to shed any light on what citizens should demand of their governments. Nevertheless, whereas Broome's "expected value theory" seeks "the correct principle for coping with uncertainty" (11), environmental pragmatists tend to focus less on debates about which general principles are the right ones for thinking about and governing human relationships with nature. Or rather, to the frustration of earnest utilitarian and Kantian applied ethicists seeking to justify crisp "hedgehog" prescriptions about how we should act and assess (which they equate with "doing ethics"), pragmatic "foxes" take the good and right to be determined experimentally, contextually, and democratically—about which more below—rather than primarily by ciphering aggregate well-being or conforming with antecedently determined moral law. If this pragmatist turn appears evasive, perhaps even happily so (West 1989), it is at any rate neither mealy-mouthed nor muddled.[7]

I should clarify that specific positions in the Breakthrough Institute's widely discussed 2011 report titled *Climate Pragmatism* (Atkinson et al. 2011) stand or fall independent of environmental pragmatism as a philosophical orientation. Their "ecomodernism"—inspired in 2003 by Nordhaus and Shellenberger, of "The Death of Environmentalism" fame—is explicitly pragmatist in its foxlike approach, but there is nothing inherent in environmental pragmatism from which one can deduce cautious support for technologies such as agricultural intensification, GMOs, and nuclear energy, or from which one can deduce whether our emphasis should be more on adaptation to climate change rather than abatement of it (see Gardiner's critique, 2011, 257).

Social Learning and Dewey's Democratic Ideal

In *Sustainable Values, Sustainable Change* (2015), Norton studies the cultural shift in the Chesapeake region of the U.S. away from object-focused "thinking like an estuary/bay" toward multi-scalar "thinking like a watershed" (250–257). This change has affected the ecological imaginations of millions, yet it did not require a victor in the prize fight over foundational environmental values. Nor did it require a prior commitment to a view about the moral standing of natural systems. However, learning to "think like a watershed" *did* minimally involve what Norton characterizes (via Kai Lee and Albert Bandura) as Deweyan "social learning." Spurred on by the work of NGOs as well as state and federal regulatory demands, the process exhibited a regional cultural shift away from narrow and short-term thinking toward longer-term, broader-range thinking. This is the sort of thinking that Norton argues can be developed as we grapple together with environmental problems. Most importantly, this shift in values occurred *through* public processes rather than as a prerequisite to participation.

Dewey argued that to be workable for social problems, we should see how far it may be practicable to conduct deliberation democratically, through the give-and-take of open dialogue and back-and-forth communication. Suspicious of being told what to do, he also criticized decision-making based on detached expert calculations of optimal welfare (e.g., 1922, MW 14:139–145), or by how forcefully you can drive home your point or sell it in the marketplace of ideas.

In Dewey's view, democratic communication maximizes the chance that we might find paths that respect legitimate interests, evaluations, and evolving identities of different individuals, institutions, and groups. When "the decider" ignores stakeholders, this raises suspicions about aims, interests, and background assumptions. It also raises issues of transparency and accountability, and it predictably leads to myopic, unworkable policies (1927, LW 2:235–372). When a decision-making process is more than nominally democratic, it seeks out tensions and divergent voices, and it gains legitimacy and direction by evaluating, criticizing, and incorporating them. Dewey consistently warned against *over*reliance on top-down, expert-driven decisions, and *where practicable* he advocated participatory processes that engage communities in social learning, fostering a public spirit of consultation to uncover troubles and to organize the expertise to deal with them. "The man who wears the shoe knows best that it pinches and where it pinches," Dewey wrote in *The Public and Its Problems* (1927), "even if the expert shoemaker is the best judge of how the trouble is to be remedied" (1927, LW 2:364).

If perceiving the need for radical changes makes one a radical, then as Dewey wrote in the middle of the Great Depression, "today any liberalism which is not also radicalism is irrelevant and doomed" (1935, LW 11:45). But Dewey's was a radicalism for grown-ups with the courage and patience to secure the "democratic means to achieve our democratic ends" (332). Or as Addams earlier made the point in her 1922 book *Peace and Bread in Time of War*: "Social advance depends as much upon the process through which it is secured as upon the result itself" (quoted in 1945, LW 15:195).

It is commonplace to note that people are overwhelmed today by momentarily exciting yet unfathomed information that is pre-prepackaged for consumption—for example, through the "echo chamber" and ideological silos of social media. As was already evident to Dewey 75 years ago, we like prepared ideas as much as we like prepared foods (1938, LW 13:95–96). In our dispersive age of globalization and connectivity, environmental pragmatists are considering with Dewey whether the disruptions and transitions we face might be better navigated with a democratic citizenry that has *also* and perhaps first learned to deal with vexing problems at a more manageable scale, beginning with our families, neighborhoods, markets, and local communities. When dealing with social matters at a local and regional scale, or via technological stand-ins for physical proximity, we may imagine more concretely and extensively the situations at hand, exchange and assess relevant information and observations, and confer about ways to mediate conflicts to converge upon solutions. Parochialism abounds, but communal ties and interactions can potentially expand and reinforce our perceptions and judgments by enabling us to draw on a cumulative wealth of experience, which is preparatory for reforming wider affairs and institutions. Dewey proposed that such communication, when scientifically informed, offers a participatory medium for awakening our slumbering democratic imaginations.

Conclusion

Returning to Jamieson's observations about harm, any adequate climate ethics must perceive and respond to a wide range of harms, the causes of which are frequently systemic, not just individual. Climate ethics must extend perception deeper into the socio-cultural, natural, and interpersonal relationships in which we are embedded. Whether this means the basic alternatives we call "morality" have failed, as Jamieson argues (2014), is an open question. What is clear is that, even amid rising global awareness of the unplanned systemic effects that radiate from our actions—climate disruption, alienated work, resource depletion, massive animal suffering, institutionalized discrimination—many retain a cultural tendency to think of themselves as "contained" in the world like a marble in a box. Concurrently, as utilitarians such as Jamieson and Singer (2015) have long argued, people too rarely imagine consequences beyond the ones in which they, or those near-and-dear, feature as the central figures. This leaves us ignorant of the environmental, social, and inter-species hazards posed by our business-as-usual behaviors.

Without taking sides on Singer's sentientist or Callicott's ecocentric positions in the winner-take-all values debate that has typified environmental and animal ethics since the 1970s, environmental pragmatists urge that we are greatly in need of wise ecological perception of the complex nature of problems, cultivated empathy for those affected by our choices, imaginative probings for technical and communal solutions, sensitivity to cultural traditions, and rich aesthetic responses to natural and cultural landscapes. We need a sort of "relational virtuosity" (Ames 2007, 55ff.) to navigate entanglements (Fesmire 2010).

There is a familiar accusation that pragmatism, as in Light's approach to climate diplomacy, is too compromising, conciliatory, and spineless to guide action, with the implication that theorists must fall back on their customary defense of antecedent principles as their overriding focus. But even as we continue to work within our varied ethical traditions, might such principles better inform decision-making if we shift or at least expand our priorities? Some priorities that may help contemporary moral and political theorizing meet wicked problems and recover from moral jetlag include: (1) Lay bare and analyze the sorts of conflicts that constantly underlie moral and political action. This includes opening communication across diverse elements of moral and political life instead of unifying what may be independent variables (Fesmire 2019); (2) Place these elements in a wider context in which norms—such as Dworkin's focus on responsibility and self-respect (2011)—gain practical traction in the entanglements of non-ideal conditions (Pappas 2019; Minteer 2018); (3) Assess the hypothesis that we can experimentally work out together what is progressive and regressive while anticipating that progress in one relevant moral dimension of a problem may be regressive in another (Kitcher 2014; Johnson 2014; Thompson 2015); and (4) Expand prospects for social learning and convergence on policy and action (Light 2017; Norton 2015; Thompson 2015; McKenna 2018; Minteer 2011).

If we are going to deal more intelligently with environmental entanglements, then we must cultivate better conditions for dialogue, debate, and persuasion. Moral fundamentalist habits, and the monistic one-way assumption that unintentionally—but not blamelessly—exercises and unduly reinforces them, are obstacles to fostering habits of moral and political inquiry better suited to responding to predicaments rapidly transforming our warming planet. A critical focus on pragmatic pluralism would be salutary.

Notes

1 Citations of John Dewey's correspondence are to *The Correspondence of John Dewey, 1871–2007*, published by the InteLex Corporation under the editorship of Larry Hickman. Citations give the date, reference number for the letter, and author followed by recipient.
2 Citations of John Dewey's works are to the thirty-seven-volume critical edition published by Southern Illinois University Press under the editorship of Jo Ann Boydston. In-text citations give the original publication date and series abbreviation, followed by volume number and page number. For example: (1934, LW 10:12) is page 12 of *Art as Experience*, which is published as volume 10 of *The Later Works*.
   Series abbreviations for *The Collected Works*: EW *The Early Works* (1882–98); MW *The Middle Works* (1899–1924); LW *The Later Works* (1925–53).
3 A public philosophy version of the moral fundamentalist pledge was published in Fesmire (2017) (first published in the Blog of the APA, April 13, 2017).
4 Cf. Anderson 2009.
5 Andrew Light, January 2016 videoconference with my Middlebury College class "The Pragmatists and Environmental Pragmatism," and March 2017 videoconference with my Green Mountain College class "Climate Justice." What follows is also drawn in part from personal communication.
6 This is perhaps ironic for a former student of Rawls to say of the father of contemporary theories of distributive justice, hence father to the main line of argument against cost-benefit analysis as the main tool for climate policy.
7 Norton's and Light's articulations of pragmatic pluralism differ somewhat from those developed by philosophers who work more overtly in the American tradition that included Dewey. McDonald's edited volume *Pragmatism and Environmentalism* (2012) offers a good representation of the latter. Cf. McKenna and Light's edited volume *Animal Pragmatism* (2004).

Bibliography

Ames, Roger T. 2007. "'The Way Is Made in the Walking': Responsibility as Relational Virtuosity," in *Responsibility*, ch. 3, ed. Barbara Darling-Smith. Lanham, MD: Lexington Books, 55ff.

Anderson, Elizabeth. 2009. "Toward a Non-Ideal, Relational Methodology for Political Philosophy." *Hypatia* 24, no. 4: 130–145.

Anderson, Elizabeth. 2013. *The Imperative of Integration*. Princeton, NJ: Princeton University Press.

Appiah, Kwame Anthony. 2017. *As If: Idealization and Ideals*. Cambridge, MA: Harvard University Press.

Atkinson, Robert D., Netra Chhetri, Joshua Freed, Isabel Galiana, Christopher Green, Steven Hayward, Jesse Jenkins, Elizabeth Malone, Ted Nordhaus, Roger Pielke Jr., Gwyn Prins, Steve Rayner, Daniel Sarewitz, and Michael Shellenberger 2011. *Climate Pragmatism: Innovation, Resilience, and No Regrets*. Breakthrough Institute and Information Technology and Innovation Foundation. https://www.itif.org/files/2011-climate-pragmatism.pdf?_ga=2.19041912.1321589747.1578504042-1849958892.1578504042; accessed January 8, 2020.

Bergman, Ronen. 2018a. *Rise and Kill First*. New York: Penguin.

Bergman, Ronen. 2018b. "Journalist Details Israel's 'Secret History' Of Targeted Assassinations" https://www.npr.org/2018/01/31/582099085/journalist-details-israels-secret-history-of-targeted-assassinations; accessed January 8, 2020.
Berlin, Isaiah. 1953. *The Hedgehog and the Fox*. New York: Simon and Schuster.
Boisvert, Raymond D., and Lisa Heldke. 2016. *Philosophers at Table*. London: Reaktion Books.
Broome, John. 2012. *Climate Matters*. New York: W.W. Norton & Company.
Callicott, J. Baird. 1999. *Beyond the Land Ethic*. Albany, NY: SUNY Press.
Caney, Simon. 2005. "Cosmopolitan Justice, Responsibility, and Global Climate Change." *Leiden Journal of International Law* 18: 747–775.
Dewey, John. 1917. "The Need for a Recovery of Philosophy." MW 10:3–48.
Dewey, John. 1922. *Human Nature and Conduct*. MW 14.
Dewey, John. 1925/1929. "Experience and Nature." LW 1.
Dewey, John. 1927. *The Public and Its Problems*. LW 2.
Dewey, John. 1929. *The Quest for Certainty*. LW 4.
Dewey, John. 1930. "Three Independent Factors in Morals." LW 5:279–288.
Dewey, John, and James H. Tufts. 1932. *Ethics*, 2nd edition. LW 7.
Dewey, John. 1935. *Liberalism and Social Action*. LW 11:1–65.
Dewey, John. 1938. "Democracy and Education in the World of Today." LW 13:294–303.
Dewey, John. 1939. *Theory of Valuation*. LW 13:189–254.
Dewey, John. 1945. "Dualism and the Split Atom: Science and Morals in the Atomic Age." LW 15:199–203.
Dworkin, Ronald. 2000. *Sovereign Virtue*. Cambridge, MA: Harvard University Press.
Dworkin, Ronald. 2011. *Justice for Hedgehogs*. Cambridge, MA: Harvard University Press.
Edenhofer, Ottmar, and Martin Kowarsh. 2015. "Cartography of Pathways: A New Model for Environmental Policy Assessments." *Environmental Science & Policy* 51: 56–64.
Elgin, Catherine Z. 1997. *Between the Absolute and the Arbitrary*. Ithaca, NY: Cornell University Press.
Fesmire, Steven. 2003. *John Dewey and Moral Imagination: Pragmatism in Ethics*. Bloomington, IN: Indiana University Press.
Fesmire, Steven. 2010. "Ecological Imagination." *Environmental Ethics* 32, no. 2:183–203.
Fesmire, Steven. 2015. *Dewey*. London and New York: Routledge.
Fesmire, Steven. 2017. "Helping Student Activists Move Past 'Us vs. Them'," *The Conversation* May 3, 2017, https://theconversation.com/helping-student-activists-move-past-us-vs-them-76838, accessed January 8, 2020.
Fesmire, Steven. 2019. "Beyond Moral Fundamentalism: John Dewey's Pragmatic Pluralism in Ethics and Politics." In Steven Fesmire (ed.), *The Oxford Handbook of Dewey*. Oxford: Oxford University Press.
Franks, Benjamin, Stuart Hanscomb, and Sean F. Johnston. 2017. *Environmental Ethics and Behavioural Change*. London: Routledge.
Gardiner, Stephen M. 2011. *A Perfect Moral Storm: The Ethical Tragedy of Climate Change*. Oxford: Oxford University Press.
Gardiner, Stephen M. 2017. "Climate Ethics in a Dark and Dangerous Time." *Ethics* 127: 430–465.
Gaus, Gerald. 2016. *The Tyranny of the Ideal: Justice in a Diverse Society*. Princeton, NJ: Princeton University Press.
Gewirth, Alan. 1978. *Reason and Morality*. Chicago: University of Chicago Press.
Greene, Joshua. 2013. *Moral Tribes*. New York: Penguin.

Hourdequin, Marion. 2014. *Environmental Ethics*. London: Bloomsbury.
James, William. 1977. *A Pluralistic Universe*. Cambridge, MA: Harvard University Press.
Jamieson, Dale. 2014. *Reason in a Dark Time: Why the Struggle Against Climate Change Failed—and What it Means for Our Future*. Oxford: Oxford University Press.
Johnson, Mark. 2014. *Morality for Humans*. Chicago: University of Chicago Press.
Jotzo, Frank. 2016. "The Climate Change Authority's Gamble on Political Pragmatism." *The Conversation*. https://theconversation.com/the-climate-change-authoritys-gamble-on-political-pragmatism-64745, accessed January 8, 2020.
Kant, Immanuel. 2002. *Critique of Practical Reason*. Trans. Werner Pluhar. Indianapolis, IN: Hackett Publishing Company.
Kitcher, Philip. 2014. *The Ethical Project*. Cambridge, MA: Harvard University Press.
Kitcher, Philip, and Evelyn Fox Keller. 2017. *The Seasons Alter*. New York: Liveright.
Latour, Bruno. 1993. *We Have Never Been Modern*. Trans. Catherine Porter. Cambridge, MA: Harvard University Press.
Light, Andrew and Eric Katz, eds. 1996. *Environmental Pragmatism*. London: Routledge.
Light, Andrew. 2013. "An Equity Hurdle in International Climate Negotiations." *Philosophy and Public Policy Quarterly* 31, no. 1: 28–35.
Light, Andrew. 2017. "Climate Diplomacy." In Stephen M. Gardiner and Allen Thompson (eds.), *The Oxford Handbook of Environmental Ethics*. Oxford: Oxford University Press.
Lovejoy, Arthur O. 1963. *The Thirteen Pragmatisms and Other Essays*. Baltimore: Johns Hopkins University Press.
McDonald, Hugh (ed.). 2012. *Pragmatism and Environmentalism*. New York and Amsterdam: Rodopi.
McKenna, Erin. 2018. *Livestock*. Athens, GA: University of Georgia Press.
McKenna, Erin, and Andrew Light (eds.). 2004. *Animal Pragmatism*. Bloomington, IN: Indiana University Press.
Mead, George Herbert. 1930. "The Philosophies of Royce, James, and Dewey in Their American Setting." In *John Dewey*. Cambridge, MA: Harvard University Press, 104–5.
Mill, John Stuart. 1986. *On Liberty*. Amherst, NY: Prometheus Books.
Mills, Charles W. 2005. "'Ideal Theory' as Ideology." *Hypatia* 20, no. 3: 165–184.
Mills, Charles W. 2017. *Black Rights/White Wrongs: The Critique of Racial Liberalism*. Oxford: Oxford University Press.
Minteer, Ben A., and Collins, J. P. 2005. "Ecological Ethics: Building a New Tool Kit for Ecologists and Biodiversity Managers." *Conservation Biology* 19: 1803–1812.
Minteer, Ben A. 2011. *Refounding Environmental Ethics*. Philadelphia: Temple University Press.
Minteer, Ben A., Jane Maienschein, and James P. Collins, eds. 2018. *The Ark and Beyond*. Chicago: University of Chicago Press.
Moellendorf, Darrel. 2014. *The Moral Challenge of Dangerous Climate Change*. Cambridge, UK: Cambridge University Press.
Neiman, Susan. 2008. *Moral Clarity*. Princeton, NJ: Princeton University Press.
Niemeyer, Simon. 2013. "Democracy and Climate Change: What Can Deliberative Democracy Contribute?" *Australian Journal of Politics and History* 59, no. 3: 430–449.
Norton, Bryan. 1991. *Toward Unity Among Environmentalists*. Oxford: Oxford University Press.
Norton, Bryan. 2005. *Sustainability: A Philosophy of Adaptive Ecosystem Management*. Chicago: University of Chicago Press.
Norton, Bryan. 2015. *Sustainable Values, Sustainable Change: A Guide to Environmental Decision Making*. Chicago: University of Chicago Press.
Nozick, Robert. 1974. *Anarchy, State, and Utopia*. New York: Basic Books.

Pappas, Gregory. 2008. *John Dewey's Ethics*. Bloomington, IN: Indiana University Press.
Pappas, Gregory. 2019. "The Starting Point of Dewey's Ethics and Sociopolitical Philosophy." In Steven Fesmire (ed.), *The Oxford Handbook of Dewey*. Oxford: Oxford University Press.
Rawls, John. 1971. *A Theory of Justice*. Cambridge, MA: Harvard University Press.
Rittell, Horst, and Melvin Webber. 1973. "Dilemmas in a General Theory of Planning." *Policy Sciences* 4: 155–169.
Sarkar, Sahotra, and Ben A. Minteer, eds. 2018. *A Sustainable Philosophy—The Work of Bryan Norton*. New York: Springer.
Sen, Amartya. 2009. *The Idea of Justice*. Cambridge, MA: Harvard University Press.
Singer, Peter. 2004. *One World: The Ethics of Globalization*. New Haven: Yale University Press.
Singer, Peter. 2015. *The Most Good You Can Do*. New Haven: Yale University Press.
Taylor, Charles. 1982. "The Diversity of Goods." In Amartya Sen and Bernard Williams (eds.), *Utilitarianism and Beyond*. Cambridge: Cambridge University Press, 129–135.
Thompson, Paul, and Kyle Whyte. 2012. "What Happens to Environmental Philosophy in a Wicked World?" *Journal of Agricultural and Environmental Ethics* 25, no. 4: 485–498.
Thompson, Paul. 2015. *From Field to Fork*. Oxford: Oxford University Press.
Valentini, Laura. 2012. "Ideal vs. Non-ideal Theory: A Conceptual Map," *Philosophy Compass* 7, no. 9: 654–664.
Watterson, Bill. 1988. *Calvin and Hobbes*. London: Sphere Books.
West, Cornel. 1989. *The American Evasion of Philosophy*. Madison, WI: University of Wisconsin Press.
Weston, Anthony. 1991. "On Callicott's Case Against Moral Pluralism." *Environmental Ethics* 13, no. 3: 283–286.
Williams, Bernard. 1985. *Ethics and the Limits of Philosophy*. Cambridge, MA: Harvard University Press.

12

PHENOMENOLOGY AND THE ETHICS OF DIFFERENCE

Levinas, Responsibility, and Climate Change

William Edelglass

Introduction: The Ethics of Difference and Singularity

The ethical dimension of difference and singularity is at the heart of several contemporary philosophical movements.[1] These movements have been influential in the humanities and social sciences, animating a critical resistance to exclusion and violence based on differences of race, class, gender, sexual orientation, religion, culture, ability, species, and other distinctions. They share a family resemblance in valuing multiplicity and rejecting the normalization of sameness and the oppression that accompanies it. Thus, they understand the ethical relationship as responding with openness and accountability to the singularity and vulnerability of the other, rather than acting according to one or another universal principle.

The ethics of difference is largely inspired by poststructuralist, postmodern, and postcolonial thinkers who valorize a difference that is not to be overcome, but instead calls for respect and recognition. In this, they consciously set themselves against what they see as the dominant tradition of Western thought, which they characterize as prioritizing unity over plurality, the one over the many, and identity over difference. According to poststructuralist, postmodern, and postcolonial critiques, the logic of identity—or universality, or totality—is inadequate and misleading. When presented as the only true and necessary account of reality, it denies its personal and historical conditions and thus seeks to hide its own situatedness and its activity in constituting concepts. And, according to these critiques, the logic of identity implies that no new experience, or question, or insight from others, will fundamentally shift the foundations of one's system. Another way to make this point is to say that these philosophers propose varieties of antifoundationalism: if, as some of them argue, difference itself is co-constitutive of meaning, ontology, and language, then there can be no ultimately stable philosophical foundations.

Many poststructuralist, postmodern, and postcolonial thinkers, then, are interested in a difference that is maintained, a difference that is precisely *not* sublimated into a unified truth. However, this interest is not just epistemological or ontological. In the wake of so much violence justified by an appeal to difference, the project of understanding difference without subsuming it to the logic of identity has become a focus of ethics. For many European philosophers of difference, the Holocaust is the manifestation of a thinking that has no tolerance for plurality; they see Nazism as the assertion of one particular identity that possesses worth, and anything that is different is subordinated, marginalized, or eliminated. In the words of the Frankfurt School philosopher Theodor Adorno: "Auschwitz confirmed the philosopheme of pure identity as death" (Adorno 2007, 362). Colonialism, patriarchy, white supremacy, and other structures that valorize and normalize one category of person to the exclusion of others can also be interpreted as the expression of "the philosopheme of pure identity," in which difference is understood in a hierarchical system that legitimates exclusion and violence.

For Adorno, and some of the other prominent figures who have contributed in their own distinct ways to philosophies of difference—including Giorgio Agamben, Judith Butler, Gilles Deleuze, Jacques Derrida, Franz Fanon, Michel Foucault, Luce Irigaray, Julia Kristeva, Emmanuel Levinas, Jean-François Lyotard, Jean-Luc Nancy, and others—any identity or characteristic of a moral or political community can always become the basis of exclusion: "all rational beings deserve moral consideration," but inevitably some humans are not regarded as "fully rational"; "all human beings deserve moral consideration," but some are not regarded as "fully human." Or, to be more explicit, some vulnerable beings are not as morally considerable because of their differences from one norm or another. For this reason, philosophies of difference that attempt to articulate and valorize a difference that is not subsumed by identity often include the development of an ethics grounded in difference.

There are so many varieties of the ethics of difference it would be impossible to characterize them all in one chapter. Here, I will focus on the philosophy of Emmanuel Levinas (1906–95), a Lithuanian-French phenomenologist. Levinas studied with Edmund Husserl and Martin Heidegger, the two most important figures in the phenomenological tradition. Levinas's accounts of language, time, work, home, ethics, justice, ontology, and other philosophical themes, are grounded in nuanced descriptions of the lived-experience of the encounter with another human. Levinas's phenomenological ethics of difference and singularity has been particularly influential, provoking increased attention to ethical aspects of social life in a variety of fields. In the following section, I present an overview of Levinas's ethics, with the aim of understanding why, as he says in an interview, "man's relationship with the other is *better* as difference than as unity" (Levinas and Kearney 1986, 22). I then devote the remainder of the chapter to exploring

Levinas's resources for climate ethics, especially his account of responsibility, his insistence on the non-reciprocal character of ethics, and his grounding of the universality of justice in the asymmetrical ethical relation.

Levinas's Phenomenology of Difference and Singularity

Levinas, as much as any thinker of difference, forcefully articulates the critique of Western philosophy as totalizing.[2] According to Levinas, rationalism, empiricism, realism, idealism, and most other philosophical movements, despite their apparent diversity, are motivated by the aspiration to encompass everything, to leave nothing exterior to thought. He finds this aspiration permeating Western philosophical traditions, from their beginnings with Socrates and his teaching of maieutics, that the soul is always already complete and contains the universe of wisdom. In these traditions, what is other is appropriated by knowledge, categorized under a concept, and—by means of philosophical methods such as dialectics or mediation or analysis—absorbed into what Levinas calls "the Same." "Western philosophy," Levinas writes, "coincides with the disclosure of the other where the other, in manifesting itself as a being, loses its alterity. From its infancy, philosophy has been struck with a horror of the other that remains other—with an insurmountable allergy" (Levinas 1986, 346). By understanding thought as possessing privileged access to the other, by approaching the other as a relationship between a knowing subject and an object of knowledge, for much of Western thought there is no other qua other. Levinas does recognize exceptions to this general tendency, even in authors such as Plato or Descartes, whose thought otherwise strips the other of singularity. Still, according to Levinas, the intellectualist project of Western philosophy can be characterized as a drive to encompass the other who becomes an object of cognition.

This totalizing drive of Western thought, according to Levinas, is not morally neutral. In "Signature," a brief autobiographical sketch, Levinas writes that his life and work were "dominated by the presentiment and the memory of the Nazi horror" (Levinas 1990, 291). Levinas himself entered the French military and was captured shortly after the start of the Second World War; he spent the next four years in a labor camp in Germany for French officers, hearing rumors of the murder of European Jews. His parents, grandparents, both his brothers, and many others close to him were killed by the Nazis; Levinas's wife and daughter spent the war years hiding in France. For Levinas, the Nazi horror and the goodness of some when confronted by overwhelming force, demand a radical rethinking of ethics, which became the central philosophical project of his life. It is primarily a project of rethinking our relationship to the other person, which, as he notes, also demands a rethinking of subjectivity.

Levinas describes the ways in which the other person is often constituted as another me, an *alter ego*, understood through empathy. Indeed, as with any other object, the meaning of the other person arises in linguistic, historical, cultural, and psychological systems. The other person can occasion satisfaction or frustration; can serve as a means or an obstacle to my ends. The other as *alter ego* can also be similar or comparable to me, and we can constitute a community under some third term to which we both belong, such as political interest, cultural practice, or geographical location. But, according to Levinas's phenomenological descriptions, beneath the community grounded in similarity, beneath the difference that is absorbed into sameness, there is a radical difference, a singular Other who is not an *alter ego*.[3] This Other cannot be synthesized with the I in a common category at a higher genus. The difference of the Other is not based on a property or characteristic, a difference that could be recognized against the background of a shared nature. Beyond the differences of characteristics, outside the differences of culture and history, irreducible to differences of physiology and psychology, Levinas describes an Other who is absolutely different, *utterly singular*.

Beginning with writings in the 1950s, and developed at length in *Totality and Infinity: An Essay on Exteriority* (1969), Levinas describes "the face" as the manifestation of the singular Other. The face-to-face relation, for Levinas, is the "ultimate situation" (Levinas 1969, 81). The face that interests Levinas is not merely an object of perception, with particular shapes and colors, or an object studied by anatomy and physiology. For Levinas the face appears as an aesthetic image, but it is ever exceeding the aesthetic, overflowing representation, and cannot be reduced, Levinas argues, even to this image of overflowing. The face is an "enigma"; that is, in Levinas's vocabulary, the face is not a phenomenon at all. Thus, Levinas is trying to describe something in addition to what we typically mean by "the face"; according to Levinas, the face "is neither seen nor touched" (Levinas 1969, 194). How, then, is the face manifest? The face appears in the very vulnerability and defenselessness of the other's body, susceptible to the pains of hunger, thirst, heat and cold, exhaustion, and the exposure to murder.

In Levinas's phenomenology, the vulnerability and irreducibility of the face "opens the primordial discourse whose first word is obligation" (Levinas 1969, 201). The face, then, is unique in that its meaning originates not in me but in the Other: the face of the Other signifies a sense not constituted by the self. This sense of the ethical arrives from beyond my world of cultural contexts and the satisfaction of needs. According to Levinas, the closest approximation in familiar linguistic terms is the command "thou shall not kill." But even this is something of a betrayal in language (Edelglass 2005). For Levinas is not describing the origin of any particular obligation. Rather, he is describing the arising of moral consciousness as such, the welcoming approach to the other in which he locates the origin of ethics.

According to Levinas, then, to attend to the face of the other who is vulnerable and exceeds my comprehending grasp is not a cognitive procedure but a welcoming approach. To attend to the vulnerability of the face is not to hold a representation of injustice, hunger, thirst, or other psychic or physical pains. It is not a vision of universal principles or an insight into a kingdom of ends. "To recognize the Other is to recognize a hunger. To recognize the Other is to give" (Levinas 1969, 75).

What should be clear from my discussion so far is that Levinas's ethics is quite different from most Western moral philosophies.[4] In what follows, I review some of the elements of his work that are particularly distinctive and that I will draw on in my exploration of a Levinasian approach to climate ethics.

First, Levinas's ethics is grounded in a phenomenological method. Thus, he approaches ethics through a description of the relation with the Other, rather than focusing on moral principles that have been abstracted from this relation. The characteristics of Levinas's ethics that I discuss below—the asymmetry of the ethical relation; a responsibility beyond intention; the way in which the ethical relation singularizes me as a subject; that ethics is first philosophy; that there is an ethics prior to universal moral prescriptions; and that ethics grounds the universality of justice—are not the result of moral reasoning based on first principles. Instead, according to Levinas, they are the results of a phenomenological inquiry into the encounter with the Other that is attentive to vulnerability and need.

Second, for Levinas, ethics is asymmetrical. According to his phenomenology of moral consciousness, I encounter the Other as transcendence, transcending my world from a height. At the same time, my own power to protect and support the Other is disclosed as inescapable; I discover myself to have resources for the Other. No matter how little I have, I recognize that I have something and am responsible to give. The power of the face, its very defenselessness, contests my autonomy to pursue my own needs. In contrast to deontology, utilitarianism, and virtue ethics, which are characterized by universal principles or shared capacities, Levinas's ethics is not grounded in a reciprocal relation of equals. Instead, in moral consciousness *I make no demands on* the Other and find myself *subject to the demand of* the Other.

Third, because of its asymmetry, Levinasian responsibility is not grounded in intention and causality. According to most moral philosophies, and our common intuitions, we are responsible for acts that we choose, especially when we understand the consequences of our choices. Causal responsibility is generally understood to be a necessary, though not sufficient, condition for moral responsibility. Thus, we tend to recognize moral responsibility for acts that are close in space and time, where we see clear connections between perpetrators and victims. However, for Levinas, ethical consciousness is precisely the sense of an infinite responsibility for the Other, even when I myself have done nothing to harm her. The more attentive I am the more I recognize my own responsibility. Thus, according to Levinas, the satisfaction of a good conscience can only arise when I lack awareness of my situation. A bad conscience is the recognition of my own limits, of my inadequate response. But, because I can never respond adequately, a bad conscience is a sign of ethical attentiveness, a recognition of a responsibility beyond my capacities to address, beyond any choices I myself have made; it is a responsibility beyond reason in which I am responsible for the Other, held to account for the Other's suffering and need.

Fourth, according to Levinas, especially in later works such as *Otherwise Than Being or Beyond Essence* (1998), the ethical relation with the Other is what singularizes us as subjects. That is, I am not a free, autonomous subject, a center of synthesizing activity—or consciousness, mind, spirit, ego, or some other abstraction—who then encounters an Other who is *there*. Rather, it is in my susceptibility to being affected by the Other that my own *here* arises. It is the ethical relation that constitutes the ground of my subjectivity in the world. This also means that for Levinas, ethics is heteronomous: my responsibility for the Other precedes my freedom to choose whether or not I want to be subject to morality.

Fifth, because the relation with the Other is the condition for any discourse, Levinas famously claims that *ethics is first philosophy*. Language, according to Levinas's analysis, is the manifestation of a world we share with others; language always bears the possibility of articulating views and values and experiences different from my own. Objectivity is made possible by shared perspectives that could be different. But the precondition of objectivity and reason and discourse is community, is the relation with others (Schuster 2015). Thus, before we begin exploring questions of metaphysics, epistemology, and ontology, or any other theoretical or practical issues, we are already in a social context, and therefore already in an ethical relationship. Signified by the face of the Other that is exterior to the order, system, and totality of my world, the ethical is prior to and thus conditions all linguistic, historical, and cultural meaning, which only arise in the encounter with the Other. This is also why moral consciousness is precisely that orientation to the Other in which I cannot produce an argument or excuse that could justify evading or even limiting my response to the need of the Other.

Sixth, Levinas's interpretive phenomenological ethics, in contrast to most Western moral philosophies, offers no universal moral principles or prescriptions. He does not seek to answer the questions that animate these traditions, such as: "what is the best human life?"; "how should I live?"; "how ought one to act?"; or "what principles of moral reason can determine what one ought to do?" He makes no attempt to derive moral principles that can be deployed to resolve dilemmas or guide our action. Indeed, he insists he is not "constructing ethics"; his goal instead is "to find its meaning" (Levinas 1985, 90). Thus, he is not interested in justifying duties, recommending the cultivation of particular virtues, or defending specific moral calculations. Even as ethics is asymmetrical and I am subject to the demand of the Other, moral consciousness is not the recognition of this or that particular obligation which I may or may not follow. Instead, Levinas understands his project as describing the very moral consciousness that precedes and motivates all moral thinking and action. In part due to his phenomenological approach, then, Levinas's thought is not in tension with virtue ethics, deontology, consequentialism, or other moral theories; it is operating, one might say, at a prior level.

Finally, even as Levinas makes no prescriptions, he argues that ethics grounds the universality of justice and concrete moral action, which themselves demand reason and justification. Thus, while he himself is not constructing an ethics in the sense of determining rules to guide action, he clearly states that "one can without doubt construct an ethics" on the basis of his thought (Levinas 1985, 90). For, according to Levinas, I never find myself in an exclusively ethical relation; other people are always already present: "The others concern me from the first" (Levinas 1998, 157). That is, the face-to-face relation never arises somehow separated from our embeddedness in multiple social relations with their own demands.

The face-to-face is prior to and conditions other social relations, with their moral dilemmas and competing obligations, including the obligation to address our own needs. In contrast to social contract theory, then, Levinas argues, justice is not the agreement of self-interested individuals who, fearing loss of life and property, seek to preserve their goods through making agreements. Instead, the ethical relation grounds justice. Making utilitarian calculations regarding the best way to provide food, clothing, medicine, education, and housing is necessary, but it is demanded by the face-to-face encounter; it is not initially motivated by rational systems or abstract reasoning, even as rational systems and abstract reasoning are necessary to ensure that basic needs are addressed. Similarly, the cultivation of virtues is important, as it allows us to respond to others, but this response is never enough when understood in the context of the ethical relation. Levinasian ethics thus eschews prescriptions, but still calls for a response, even if this response can never be adequate. It has what Diane Perpich calls a "*normativity without norms*" (Perpich 2008, 126). That is, Levinas's phenomenology describes a normative force to respond to the Other, to engage in the world, but there is no particular rule to follow or virtue to cultivate that would be universally applicable. Hence, there is no escape from the moral dilemmas of social life, for we are incessantly called by a responsibility to address needs that always demand more than the capacities and resources we possess.

Levinas's Ethics and Climate Change

Levinas, who died in 1995, never addressed climate change in his work. But scholars have been inspired by his thought to explore "Levinasian" approaches to themes that he himself did not address, or did not address adequately, including gender, race, colonialism and its consequences, animals, and environmental thought (Chanter 2001; Atterton and Calarco 2010; Edelglass et al. 2012; Drabinski 2013). In what follows, I explore several ways in which Levinas's interpretive phenomenology, especially his description of responsibility, can be a fruitful resource for understanding the ethical dimension of climate change.

A number of scholars, most famously Dale Jamieson, have argued that climate change and other collective action environmental problems pose a challenge to our traditional conceptions of moral responsibility. Because individually my actions are inconsequential, I intend no harm, and there is no immediate victim, according to the standard account of responsibility, I am not responsible for the suffering that results from climate change. But Levinas's account of responsibility, due to its differences from the standard account, suggests that I am indeed responsible for those who suffer from a changing climate. Their vulnerability and suffering disclose my own resources to work for mitigation and adaptation, and my complicity in their situation. Moreover, Levinas's understanding of the work of ethics as a service that is non-reciprocal, a working for a future time in which I do not benefit, provides a framework for conceptualizing my responsibility to future generations, especially when the future is uncertain. Finally, grounding justice in the ethical relation, as Levinas does, motivates collective agreements aimed at mitigating and adapting to climate change while simultaneously resisting the ways in which these procedures inevitably justify present and future suffering.[5]

Climate Change as a Challenge to Ethics

Before climate change had registered beyond a small group of scientists, Hans Jonas argued that we need to rethink our understanding of responsibility in the face of the disastrous environmental and human consequences of our daily lives. According to Jonas, in *The Imperative of Responsibility: In Search of an Ethics for the Technological Age*, greater than the threat of atomic holocaust—which, he suggests, can be resolved through wise diplomacy[6]—are the threats of our everyday actions, the threats posed by "the peaceful and constructive use of worldwide technological power, a use in which all of us collaborate as captive beneficiaries through rising production, consumption, and sheer population growth" (Jonas 1984, ix). Using a language that is familiar in contemporary climate discourse, Jonas writes: "Thresholds may be reached in one direction or another, points of no return, where processes initiated by us will run away from us on their own momentum—and toward disaster" (Jonas 1984, ix). According to Jonas's analysis, the advent of modern technology, with its far-reaching impact on ecosystems, societies, and future human existence, has changed the very nature of human action. The ethical principles we have inherited work well for actions that are voluntary—in the sense that one could reasonably be expected to act otherwise—and have consequences that are relatively close in space and time. But the inadequacy of inherited ethical principles is illustrated by our responses to the environmental challenges we face today, challenges which for many seem to be purely technical problems located outside the moral realm. Jonas thus argues we need a new account of moral responsibility.

Jamieson has also been making the argument that our inherited ethical principles and intuitions are inadequate for responding to contemporary environmental challenges—especially climate change. That is, *climate change is not just an ethical challenge in the realm of action, but it also poses a challenge to our ethics as such*, especially with regard to moral responsibility. According to Jamieson, climate change "lacks some of the characteristics of a paradigm moral problem" (Jamieson 2007, 475). Why is this? For Jamieson, a paradigm moral problem—the kind of moral problem with which much of the history of moral philosophy has been concerned—is characterized by a moral agent who could have acted otherwise, who intentionally harms another. Thus there are well-defined actors who intend harm and distinct victims who suffer. Consider, as an example of a paradigm moral problem, the case of an agent who poisons a neighbor's well, depriving the neighbor of access to drinking water. In this paradigm case, the perpetrator and the victim, and how the intended actions of the one causally relate to the harm suffered by the other—and therefore the moral responsibility—can be determined with relative ease.

To illustrate the ways in which climate change does not conform to the standard account of moral responsibility, Jamieson asks us to think of variations of the paradigm case that obscure the moral dimension of the action, even as the victim continues to suffer the same harmful consequence of not having access to clean drinking water.[7] For example, what if the agent happens to be one of many individuals, who, without coordinating their actions, each toss an object into the well that on its own would be benign but cumulatively poison the well? Or, perhaps the agent and the victim are geographically distant, but the agent is one of many people who order a product made at a factory, and the factory emits pollution which poisons the victim's well. Or the agent lives several generations before the victim, and the production of goods purchased by the agent, together with many others, causes widespread degradation of water systems such that the victim is unable to access clean water.

Jamieson's point is that our moral intuition suggests that while in the first case, where the agent intentionally poisons the neighbor's well, the moral responsibility is obvious, moral responsibility becomes harder to identify with each succeeding case. There may still be agents who act and victims who suffer the very same consequences of those actions, but other characteristics of the paradigm moral problem are notably absent. For example, because of the lack of temporal or spatial proximity, causal connections between the original action and the harm that will be suffered are obscure, there is no intention to harm, and victims and perpetrators are hard to identify (Jamieson 2007, 476). As Jamieson notes in a recent paper, "Everything influences everything else in the Anthropocene, but responsibilities are elusive" (Jamieson and Di Paola 2016, 269). With no intentional perpetrators and unknown victims, it is tempting to think of climate change solely as an unfortunate problem that will be resolved by well-informed and well-intentioned public policy, prudent diplomats, scientific developments, and new technologies.

It was precisely because individual actions are inconsequential in collective action environmental problems that Hans Jonas, when he introduced a new moral imperative for contemporary industrialized society, addressed it not to the individual making choices among various actions but to those making public policy (Jonas 1984, 11). Baylor Johnson and Walter Sinnott-Armstrong have argued in a similar vein that there is no individual obligation to cut personal greenhouse gas (GHG) emissions; the only obligation we have as individuals is to support efforts to form binding collective agreements, for example by working to elect politicians who will enact legislation aimed at large-scale reductions in emissions. For Johnson, the atmosphere is a commons in need of more regulation and the problem of climate change has the structure of a tragedy of the commons. Thus, he argues, while there is a moral obligation to contribute to a collective agreement to regulate the commons, there is no unilateral obligation to reduce one's individual emissions (Johnson 2003).[8] According to this account, climate change and other common-pool resource problems constitute a kind of prisoner's dilemma, in which individuals acting according to their preferences worsen the situation for everyone. What would be required, then, is an enforceable, collective agreement.

Sinnott-Armstrong arrives at a similar conclusion by arguing that behaviors that cause more GHG emissions than relatively easy to achieve alternatives—think of frequent vacation flights; driving a gas-guzzling vehicle simply for pleasure; keeping the thermostat above 70°F in the winter—are inconsequential and thus we, as individuals, are not in fact responsible for climate change. As he writes, "No storms or droughts or heat waves can be traced to my individual act of driving" (Sinnott-Armstrong 2005, 291). For Sinnott-Armstrong, then, because these higher GHG emitting individual actions do not actually cause climate change, there is nothing morally wrong with them. The only moral obligation we have is to support systemic changes at a policy level by electing and supporting politicians who will implement incentives and regulations that will make a *consequential* difference.

While Johnson and Sinnott-Armstrong are certainly correct that policy change at national and international levels is necessary for substantive decreases in GHG emissions, it is hard to see how the obligation for individuals to work in support of such a policy change is not also a problem of collective action. Just as my driving less or persuading others to hang-dry their laundry alone will not in fact save the Himalayan glaciers or the Greenland ice sheet, so my voting for a particular candidate who favors limiting GHG emissions, or persuading others to vote for such a candidate, will also not cause the implementation of appropriate policies or prevent climate change. For the same reasons, no activism, public policy work, scientific discovery, or any other contribution to the global movement working for a decrease in GHG emissions, strictly speaking, is of any consequence. Completing their lines of argument leads to the conclusion that nothing we as individuals do causes climate change, there is nothing we as individuals can do to prevent it, and therefore there is no individual obligation or responsibility to do anything about it.

Levinasian Ethics, Individual Responsibility, and Climate Change

Climate change is both a moral challenge and a challenge to moral theory because we find ourselves caught in a situation where together we cause immense suffering and our inherited moral theory has difficulty conceptualizing any moral responsibility. Moreover, as Jamieson argues, "because we tend not to see climate change as a moral problem, it does not motivate us to act with the urgency characteristic of our responses to moral challenges" (Jamieson 2007a, 477).[9] But precisely because of the ways in which it is distinct from traditional Western moral theories, Levinas's ethics provides resources for understanding the ethical dimension of climate change.

For Levinas, our responsibility exceeds our capacity to act and we are responsible even beyond what we choose. This understanding departs significantly from what Jamieson describes as the account of responsibility implicit in "paradigm moral problems." As Perpich points out, Levinas inverts "the standard account" of responsibility according to which we are morally responsible only for actions that are within our capacity and voluntary (Perpich 2008, 78–123). In the standard account, because *ought* implies *can*, responsibility does not demand more than is reasonably possible for the agent. And while moral philosophers generally agree that we ought not pursue our own interests when this involves severe detriment to others, as these others become more distant, according to the standard account, our responsibility diminishes. In the standard account, then, Perpich notes, "there is indeed a limit to that for which I can be held responsible and this limit is generally determined through a consideration of the proximity of the agent's action to the matter or affair in question" (Perpich 2008, 82). The Levinasian inversion of the standard account, however, means that I am responsible for the other regardless of how inconsequential my actions might be.

Levinasian responsibility is not grounded in intention, or immediate causal connections between my actions and the pains of others. Rather, Levinasian responsibility arises with an attention to the vulnerability and suffering of others whose well-being and existence is threatened, a suffering that disrupts, so Levinas argues, all justifications for inaction (Edelglass 2006). As Y.A. Kang observes of Levinas's thought: the cry of suffering discloses "not one of several themes which could be approached from an ethical perspective. Suffering is precisely the opening of the ethical perspective" (Kang 1997, 498). The vulnerability of the other, the other's susceptibility to pain and damage, according to Levinas, disorients consciousness with an appeal. This appeal is not based on the other's greater possession of power, but is grounded in vulnerability, in the other's lack of resources. This is how responsibility is first disclosed. If I am attentive to the suffering that arises from climate catastrophe, then, regardless of whether my own actions are causally inconsequential, I am aware of my responsibility.

A Levinasian attention to responsibility for the suffering Other might also dispute the primary basis of Sinnott-Armstrong and Johnson's arguments, namely, that as individuals we do not cause harm to others when we drive our cars, heat our homes with fossil fuels, and cause GHG emissions in meeting our daily needs and desires. The more attentive I am, the more I recognize that nourishing myself requires taking resources from the Other, that in caring for my daily needs I inflict a violence on the Other. Thus, for Levinas: "the important question of the meaning of being is … do I not kill by being?" (Levinas 1985, 10). As Levinas noted, when we eat, we take food that might nourish an Other. Additionally, today we might say, when we choose actions that cause GHG emissions, similarly, we are harming others. Levinas insists we "cannot live without killing, or at least without taking the preliminary steps for the death of someone" (Levinas 1985, 10). Thus, Levinas's account of responsibility is true to the fact that even as we might cultivate many of the green virtues—and reduce our emissions by driving a hybrid; or taking public transportation; or riding a bicycle; drawing our electricity from renewable sources; growing much of our own food; and successfully engaging in our local institutions and politics to enact more sustainable policies and practices—still, we are complicit in contributing to suffering on an unimaginably vast scale.

To live off the resources of industrialized society in the Anthropocene, then, is to be complicit in present and future suffering. From a Levinasian perspective, the more attentive I am, the more I feel responsible for shortages of food and fresh water due to desertification, droughts, storms, erosion, and flooding. And for the dramatically increased rate of collapse and extinction of our fellow species.[10] And for the increased rates of death, disease, migration, economic loss, fires, more powerful storms, and more intense heat waves that researchers in the natural and social sciences have already documented and attributed to anthropogenic climate change.

As Hans Jonas argued, we must recognize how, because it is correlated with the new powers that we exercise, responsibility has dramatically increased: "The lengthened reach of our deeds," Jonas concludes, "moves *responsibility*, with no less than man's fate for its object, into the center of the ethical state" (Jonas 1984, x). According to Jonas, this also moves responsibility from the individual to policy makers and legislators. But, as I have argued, Levinas offers an account that makes sense of the responsibility many of us feel as individuals in the Anthropocene. Moreover, as I shall address in the next section, this responsibility is oriented to both present and future generations, even when that future is uncertain.

Levinasian Responsibility, Future Generations, and the Uncertain Future

For Jonas, the new character of human action requires an ethic that takes account of the extended future. Thus, he argues for a new moral imperative. According to Jonas, the imperative today is:

> "Act so that the effects of your action are compatible with the permanence of genuine human life"; or expressed negatively: "Act so that the effects of your action are not destructive of the future possibility of such life"; or simply: "Do not compromise the conditions for an indefinite continuation of humanity on earth"; or, again turned positive: "In your present choices, include the future wholeness of Man among the objects of your will."
>
> *Jonas 1984, 11*

Jonas wrote these words addressing environmental challenges before climate change was widely understood. But climate change demands attention to the needs of future generations of humans and our fellow creatures, and the ecosystems that will support them. Yet as Jamieson points out in his discussion of the "paradigm moral problem," one of the difficulties in thinking about moral responsibility and climate change is the lack of spatial and temporal proximity between agents and victims. One might expect this to be a problem for Levinas as well, whose accounts of responsibility include descriptions of a face-to-face encounter and a relation to the Other in proximity. But for Levinas, the ethical is precisely that which is nonreciprocal, a departure without return, a one-way movement. And a work for the distant, especially for the future, characterized by a horizon that exceeds my own time, is then characteristic of the ethical *par excellence*.

Thus, while the ethical arises in proximity, for Levinas, it also makes a demand for that which is beyond us in space and time. Indeed, he characterizes ethics as "work without remuneration, whose result is not allowed for in the time of the agent, and is assured only for patience, a work that is effected in the complete domination of and surpassing of my time" (Levinas 1996, 50). Ethics is precisely not self-reflexive; it is a service undertaken without hope of reciprocity. It is Levinas's "to-be-for-after-my-death" (Levinas 1986, 349), in the place of Heidegger's "being-towards-death." The ethical is precisely an orientation beyond myself, which Levinas describes as a future without me.

Thus, a Levinas-inspired approach to climate change would be directly in opposition to economic analyses based on a future discount rate. These analyses, derived from the principle that future costs are preferred over present costs, indicate that it is more prudent to invest funds today than to employ them for mitigating climate change; the increased capital could then be employed in the future for greater adaptation. While these analyses seem blind to the science, which suggests that it is more prudent to devote resources to mitigation efforts, from a Levinasian perspective they are also lacking in moral consciousness. In contrast to economic thinking, which privileges the present in which I myself benefit, moral consciousness privileges a time beyond me, in which I cannot reap the benefit of my actions. Levinas approvingly quotes Léon Blum—writing in Bourassol prison in December 1941—quoting Nietzsche: "Let the future and the things most remote be the rule of all the present days!" (Levinas 1996, 50–51). While Levinas cites this line addressing a very different context—namely the "hole in history" that was late 1941—it serves as a suitable call when faced with anthropogenic climate change.

Levinas's emphasis on the ethical as working for a time beyond me, lends itself to an account of responsibility for future generations and spatially distant others, those who will suffer the most severe consequences of climate change. It also seems an appropriate way to approach the uncertainty of the future in a changing climate. As Stefan Skrimshire has pointed out, the discourse of "tipping points" and "points of no return" and the science underlying these apocalyptic modes of framing our situation, introduce a radical uncertainty into our ethical lives. I am called to act even as there is no reason to have confidence that my actions will have the intended impact, as there is cause to believe that the world will change in significant ways. According to Skrimshire,

> The reason many today take action on climate change has more to do with acts of commitment to the future without which the ethical status of the present would be diminished, than with the avoidance of an *excess* harm through one's actions.
>
> *Skrimshire 2009, 3*

Skrimshire is describing a situation in which we are called to a responsibility that does not depend on knowing the consequences of one's actions. Indeed, we cannot know the consequences because the future is so uncertain. Responsibility regardless of the consequences may be characteristic of deontological or virtue approaches to ethics. But it also is characteristic of a Levinasian approach, in which one is responsible beyond reason, or rather, one finds oneself responsible prior to reason. It doesn't matter if my actions will be adequate; I am still responsible and called to act, no matter how uncertain the future may be.

Levinas, Concrete Action, and Collective Agreements

In the preceding sections I have suggested that Levinas's ethics provides resources for a response to the challenge of inconsequentialism. That is, even if the consequences of my own actions are inconsequential, I am still responsible to others, both present and future generations, for the sufferings that result from climate change. To be morally attentive is to recognize that I am nourished by resources needed by others now and in the future. In short, Levinas's work can help us make sense of our responsibility as individuals in the Anthropocene. However, Levinas's work does not, in any particular instance, tell us what we ought to do. Some critics even regard his account as precluding "a genuine engagement with 'concrete reality'" (Lee 1999, 254). And indeed, Levinas's descriptions of the meaning of ethics are not motivated by a concern to derive prescriptions; his project is not a search for moral standards and *a fortiori*, he makes no attempt in his philosophical writings to apply his ethics to particular contexts. Indeed, he suggests, while I am morally responsible to offer what I have for the other, to demand universality "would be to preach human sacrifice" (Levinas 1998, 94). Levinas's emphasis on the fundamentally asymmetrical character of ethics seems to necessarily undermine any search for universally applicable moral standards or justice.

However, it is precisely the resources Levinas provides for new ways of seeing the ethical dimension of the situations in which we find ourselves, both personally and politically, that have made his work so influential across the humanities and social sciences. As Robert Bernasconi suggests, concluding a discussion of

Levinas's project of rethinking subjectivity and responsibility, "it matters not at all unless it impacts on our approach to concrete situations so that we come to see them as ethical" (Bernasconi 2002, 250). And, Perpich observes, "it is well-nigh impossible" to interpret Levinas's works "as having nothing to do with normative ethical concerns" (Perpich 2008, 125). This does not mean, she is careful to insist, that one can derive rules from Levinas's work. Rather, she argues, Levinas gives an account of responsibility such that there is a normative force to respond to the need of the Other while providing no specific norms. In an interview, Levinas himself said that one could construct a concrete ethics on the basis of his work, but that was not his project (Levinas 1985, 90). He seems to identify more closely with Kant than any other moral philosopher. But a Levinasian approach to concrete action to mitigate and adapt to climate change might very well be pluralistic. That is, it might involve the cultivation of green virtues, and also at times, the application of deontological or utilitarian principles. We are constantly finding ourselves responsible in all sorts of ways, called in many different directions, and different moral theories may be more or less helpful in one situation or another. But no moral rule will be universally appropriate for responding to the suffering of others. A Levinasian contribution to climate ethics, then, will insist on our underlying responsibility for present and future others, even as no particular action, no matter how much it conforms to some moral principle, can justify a good conscience.

According to Levinas, it is precisely the infinite and asymmetrical responsibility that demands the cultivation of virtues and respect for moral laws, that informs action in concrete situations. Indeed, Levinas's understanding of an infinite responsibility constitutive of the ethical situation is accompanied by the argument that one is never in an exclusively ethical relation. That is, I am never in an exclusively face-to-face situation with *one* Other; there are always other people who also concern me. Levinas refers to these other others as "the third," who interrupt the asymmetry of the ethical relation. In contrast to the face-to-face encounter, life with other others demands that I recognize rules and reason, and that I recognize the need to take care of myself as one among others. The asymmetry of ethics, with its infinite responsibility, then, is not the contestation of normativity and rules but, according to Levinas, its grounding. It is the infinite responsibility that demands both virtuous action and adherence to moral rules, and, on the political level, the just organization of society. Against the view that the source of justice is self-interest, then, Levinas argues that the ethical relation is the source of both justice and morality.

Clearly, as Hans Jonas, Baylor Johnson, Walter Sinnott-Armstrong, Stephen Gardiner, and many others argue, public policy, legislation, and international political agreements *are* necessary to make significant progress in mitigating and adapting to climate change. Levinas does not provide resources or guidance regarding whether the international community should set up a climate regime according to equal per capita entitlements, rights to subsistence emissions, priority to the least well-off, or equalizing marginal costs.[11] Nor does Levinas's account help us judge what particular funding mechanisms or political structures will best support international climate agreements. His work does not provide specific rules or political and economic principles to help us adjudicate between competing proposals for the distribution of costs and benefits of mitigation, adaption, and unprevented harms due to climate change. What does a Levinasian approach to climate ethics provide? A Levinasian approach would demand an attention to those who are and will be vulnerable and suffering due to climate change and that we respond accordingly. This response might include, for example, working to ensure that the international community agrees to limiting the rise in global temperatures to 1.5 degrees Celsius. But, from a Levinasian perspective, even agreeing to this amount of warming, arrived at by weighing economic cost and political viability, will still involve rationalizing someone else's suffering. If we are morally attentive, we will still be disturbed, recognizing a responsibility to mitigate and adapt to climate change beyond our capacities.

Conclusion

A generation ago, it seems, many environmental philosophers and activists tended to ground their work in one form or another of non-anthropocentrism. These views recognized an intrinsic value in nature that made nature morally considerable. The idea that "nature," as some abstract, ahistorical other to humans and human culture could be a source of moral obligation is alien to Levinas's thought. For Levinas, "nature" is precisely that ontological realm which is outside the realm of ethics; it is distinguished instead by its drive to persist and its inability to put the Other before the self (Toadvine 2012, Nelson 2012). Today, though, many environmental thinkers and activists articulate their concerns using conceptual frameworks that are less reliant on non-anthropocentric metaphysical and ethical views. Indeed, the *climate* movement, including the intellectual work that motivates it, is now very much a *climate justice* movement. A Levinasian approach has considerable resources to contribute to thinking about climate justice. Indeed, while Levinas may not have much to say in support of an ethics of respect for nature, a Levinasian understanding of the vulnerability of humans and other animals can still justify caring for the earthly elements that we enjoy and that nourish humans and other animals (Perpich 2012; Benso 2012).

In the context of anthropogenic climate change, a phenomenology of moral consciousness discloses my own resources to respond to this suffering. Moral consciousness also reveals my resources for working toward mitigation and adaptation to minimize harm to future generations of humans and other others. What would a Levinas-inspired approach to climate change look like in practice? Perhaps it would appear similar to Jamieson's environmental virtues, especially temperance (living moderately, consuming less) and mindfulness (the awareness of how one's actions radiate out across space and time). But because Levinas's ethics includes a demand for justice, a Levinas-inspired approach to climate change will—in accord with Johnson, Sinnott-Armstrong, and Gardiner—also be a call to work for the kinds of regulation, legislation, and collective agreements that can result in systemic reductions in GHGs. Thus, I am called to respond in multiple ways to the vulnerability and unimaginably vast suffering that comes with climate change. Levinas's recurrent concern with "the precariousness of the Other" (Levinas 1996, 167) resonates with the contemporary emphasis on climate *justice*. Indeed, his ethics of difference and singularity provides an apt description of ethical life in the Anthropocene: I have inescapable responsibilities that increase with my awareness and attention and are always beyond my capacities to meet.

Notes

1 I am grateful to Ted Toadvine, Bronwen Tate, and Martin Shuster for their helpful comments on a previous draft of this chapter.

2 Some of the passages in this section were originally written for a course on ethics for MA students in philosophy at the University of Mumbai and are used with permission.

3 Levinas generally uses *Autrui* to refer to the other human, the absolutely other, and *autre* to refer to the other that is an object of consciousness. In this chapter, "Other" translates *Autrui*, the absolutely other; "other" translates *autre*.

4 For texts that engage Levinas's work in conversation with classical and contemporary moral philosophy, see Diane Perpich, *The Ethics of Emmanuel Levinas* (2008), and Michael L. Morgan, *Discovering Levinas* (2007).

5 Some of the passages that follow originally appeared in Edelglass 2012 and are used with permission.

6 From the perspective of 2020, we may not be as confident as Jonas was in the 1970s, that wise diplomacy could resolve the threat of nuclear holocaust.

7 I have altered Jamieson's metaphor here. He uses an example of Jack stealing Jill's bicycle as a paradigm moral problem (Jamieson 2007, 476). In earlier work, he had made the same point about the inadequacy of inherited moral principles: they work well when applied to someone breaking into a home to steal a television but not so well when applied to collective action problems such as climate change (Jamieson 1992, 148). Jamieson also discusses the series of examples of the loss of a bicycle to illustrate the inadequacy of traditional principles and intuitions of moral responsibility in *Reason in a Dark Time* (2014). There is some debate regarding how apt the loss of a bicycle is to the harms caused by climate change. Stephen M. Gardiner argues that it is a misleading example and alters it somewhat to argue that it is a failure not of our concept of moral responsibility, but of our political institutions (Gardiner 2011). For a discussion of these competing examples, and a proposal for why our moral principles are not as problematic as Jamieson supposes, see Kingston (2014).

8 More recently, Johnson has suggested that there is indeed an individual obligation to reduce personal emissions, but that it is secondary to the obligation to work towards public structures that will enable larger-scale reductions (Johnson 2011).

9 Jamieson himself has suggested that we can find this moral urgency in recognizing a duty to respect nature (Jamieson 2010, 440–43). In earlier writings he had proposed virtue theory as the best way to think about morality in the context of climate change. According to Jamieson, even a strict utilitarian should be a virtue theorist when it comes to analyzing the moral dimension of climate change, as cultivating green virtues is the best possible way of achieving the greatest good for the greatest number (Jamieson 2007b).

10 For an account of the limitations and resources for thinking about our ethical obligations to other animals from a Levinasian perspective, see Atterton 2012.

11 For a discussion of these four prominent proposals, see Gardiner (2004).

Bibliography

Adorno, Theodor W. 2007. *Negative Dialectics*. Translated by E. B. Ashton. New York: Continuum.

Atterton, Peter, and Matthew Calarco. 2010. *Radicalizing Levinas*. Albany: State University of New York Press.

Atterton, Peter. 2012. "Facing Animals." In *Facing Nature: Levinas and Environmental Thought*, edited by William Edelglass, James Hatley, and Christian Diehm, 25–39. Pittsburgh: Duquesne University Press.

Benso, Silvia. 2012. "Earthly Morality and the Other: From Levinas to Environmental Sustainability." In *Facing Nature: Levinas and Environmental Thought*, edited by William Edelglass, James Hatley, and Christian Diehm, 191–208. Pittsburgh: Duquesne University Press.

Bernasconi, Robert. 2002. "To Which Question is 'Substitution' the Answer?" In *The Cambridge Companion to Levinas*, edited by Simon Critchley and Robert Bernasconi, 234–251. Cambridge: Cambridge University Press.

Chanter, Tina, ed. 2001. *Feminist Interpretations of Emmanuel Levinas*. University Park: Pennsylvania State University Press.

Drabinski, John E. 2013. *Levinas and the Postcolonial: Race, Nation, Other*. Edinburgh: Edinburgh University Press.

Edelglass, William. 2005. "Levinas's Language." *Analecta Husserliana* 85: 47–62.

Edelglass, William. 2006. "Levinas on Suffering and Compassion." *Sophia* 45, no. 2: 43–59.

Edelglass, William. 2012. "Rethinking Responsibility in an Age of Anthropogenic Climate Catastrophe." In *Facing Nature: Levinas and Environmental Thought*, edited by William Edelglass, James Hatley, and Christian Diehm, 209–228. Pittsburgh: Duquesne University Press.

Edelglass, William, James Hatley, and Christian Diehm, eds. 2012. *Facing Nature: Levinas and Environmental Thought*. Pittsburgh: Duquesne University Press.

Gardiner, Stephen M. 2004. "Ethics and Global Climate Change." *Ethics* 114: 555–600.

Gardiner, Stephen M. 2011. "Is No One Responsible for Global Environmental Tragedy? Climate Change as a Challenge to our Ethical Concepts." In *The Ethics of Global Climate Change*, edited by D.G. Arnold, 38–59. Cambridge: Cambridge University Press.

Jamieson, Dale W. 1992. "Ethics, Public Policy, and Global Warming." *Science, Technology, & Human Values* 17, no. 2 (Spring): 139–153.

Jamieson, Dale W. 2007a. "The Moral and Political Challenges of Climate Change." In *Creating a Climate for Change: Communicating Climate Change and Facilitating Social*

Change, edited by Susanne C. Moser and Lisa Dilling, 475–83. New York: Cambridge University Press.

Jamieson, Dale W. 2007b. "When Utilitarians Should Be Virtue Theorists." *Utilitas* 19: 160–83.

Jamieson, Dale W. 2010. "Climate Change, Responsibility, and Justice." *Science and Engineering Ethics* 16: 431–45.

Jamieson, Dale W. 2014. *Reason in a Dark Time: Why the Struggle Against Climate Change Failed and What it Means for our Future*. New York: Oxford University Press.

Jamieson, Dale W. and Marcello Di Paola. 2016. "Political Theory for the Anthropocene." In *Global Political Theory*, edited by David Held and Pietro Maffettone, 254–279. Cambridge: Polity.

Johnson, Baylor. 2003. "Ethical Obligations in a Tragedy of the Commons." *Environmental Values* 12: 271–287.

Johnson, Baylor. 2011. "The Possibility of a Joint Communiqué: My Response to Hourdequin." *Environmental Values* 20: 147–156.

Jonas, Hans. 1984. *The Imperative of Responsibility: In Search of an Ethics for the Technological Age*. Translated by Hans Jonas with the collaboration of David Herr. Chicago: University of Chicago Press.

Kang, Y.A. 1997. "Levinas on Suffering and Solidarity." *Tijdshrift voor Filosofie* 59: 482–504.

Kingston, Ewan. 2014. "Climate Change as a Three-Part Ethical Problem: A Response to Jamieson and Gardiner." *Science and Engineering Ethics* 20, no. 4: 1129–1148.

Lee, Jung H. 1999. "Neither Totality nor Infinity: Suffering the Other." *Journal of Religion* 79: 250–279.

Levinas, Emmanuel. 1969. *Totality and Infinity: An Essay on Exteriority*. Translated by Alphonso Lingis. Pittsburgh: Duquesne University Press.

Levinas, Emmanuel. 1985. *Ethics and Infinity: Conversations with Philippe Nemo*. Translated by Richard Cohen. Pittsburgh: Duquesne University Press.

Levinas, Emmanuel. 1986. "The Trace of the Other." Translated by Richard Cohen. In *Deconstruction in Context: Literature and Philosophy*, edited by Mark Taylor, 345–359. Chicago: University of Chicago Press.

Levinas, Emmanuel. 1990. *Difficult Freedom: Essays on Judaism*. Translated by Seán Hand. Baltimore: Johns Hopkins University Press.

Levinas, Emmanuel. 1996. *Basic Philosophical Writings*. Edited by Adrian Peperzak, Simon Critchley, and Robert Bernasconi. Bloomington: Indiana University Press.

Levinas, Emmanuel. 1998. *Otherwise Than Being or Beyond Essence*. Translated by Alphonso Lingis. Pittsburgh: Duquesne University Press.

Levinas, Emmanuel and Richard Kearney. 1986. "Dialogue with Emmanuel Levinas." In *Face to Face with Levinas*, edited by Richard Cohen, 13–33. Albany: State University of New York Press.

Morgan, Michael L. 2007. *Discovering Levinas*. New York: Cambridge University Press.

Nelson, Eric Sean. 2012. "Levinas and Adorno: Can There Be an Ethics of Nature?" In *Facing Nature: Levinas and Environmental Thought*, edited by William Edelglass, James Hatley, and Christian Diehm, 109–133. Pittsburgh: Duquesne University Press.

Perpich, Diane. 2008. *The Ethics of Emmanuel Levinas*. Stanford: Stanford University Press.

Perpich, Diane. 2012. "Scarce Resources? Levinas, Animals, and the Environment." In *Facing Nature: Levinas and Environmental Thought*, edited by William Edelglass, James Hatley, and Christian Diehm, 67–94. Pittsburgh: Duquesne University Press.

Shuster, Martin. 2015. "On the Ethical Basis of Language: Some Themes in Davidson, Cavell, and Levinas." *Journal for Cultural and Religious Theory* 14, no. 2: 241–266.

Sinnott-Armstrong, Walter. 2005. "It's Not *My* Fault: Global Warming and Individual Moral Obligations." In *Perspectives on Climate Change: Science, Economics, Politics, Ethics*, edited by Walter Sinnott-Armstrong and Richard B. Howarth, 285–307. Amsterdam: Elsevier.

Skrimshire, Stefan. 2009. "Points of No Return: Climate Change and the Ethics of Uncertainty." *Environmental Philosophy* 6, no. 2: 1–20.

Toadvine, Ted. 2012. "Enjoyment and Its Discontents: On Separation from Nature in Levinas." In *Facing Nature: Levinas and Environmental Thought*, edited by William Edelglass, James Hatley, and Christian Diehm, 161–189. Pittsburgh: Duquesne University Press.

INDEX

Note: Page numbers in *italic* refer to Figures; Page numbers in **bold** refer to Tables

For Product Safety Concerns and Information please contact our EU representative GPSR@taylorandfrancis.com
Taylor & Francis Verlag GmbH, Kaufingerstraße 24, 80331 München, Germany

www.ingramcontent.com/pod-product-compliance
Ingram Content Group UK Ltd.
Pitfield, Milton Keynes, MK11 3LW, UK
UKHW022002180926
475701UK00003B/148
9781138678279